Plasma Versus the Big Bang

by Rolf A. F. Witzsche

Contents

About the Illustrated Science series
On the Ice Age and Climate Change
and the book

Plasma versus the Big Bang

Book 2 of the series: 'Cold' Plasma Fusion Powers the Sun

In the Big-Bang-cosmos mythology everything that exists in the universe exists in atomic form. A giant explosion is deemed to have created all the matter in the universe. Every sun in every galaxy is deemed to be atomic in nature, powered by fusing its hydrogen atoms into helium, whereby it burns itself out. That's the theory. The reality is the opposite of it. In the real universe, according to evidence, almost nothing exists as atomic matter. Every sun in every galaxy is a sphere of plasma, which is powered by electric interaction at its surface, with cosmic plasma that flows through the galaxies.

The Plasma Universe is not entropic in nature, but is self-creating and self-sustaining by the nature of space that is energy itself; which is a concept pioneered by David Bohm, whom Einstein had referred to as his successor.

In the real universe, nothing is running down towards an eventual energy depletion death. In fact, the universe is anti-entropic and expanding and progressing. We, ourselves are evidence of this progression. Neither is our Sun isolated from the progressive nature of the universe, but expresses its dynamics, its resonating plasma streams, and their reflection in the climate on Earth. Climate Change reflects the nature of the universe. The Earth itself is the creation of the Sun, with its atoms having been massively synthesized in high-energy times near the center of the galaxy.

The synthesizing plasma fusion is presently at a low state, though it is currently enhanced for our Sun by electromagnetic 'Primer Fields' that focus interstellar plasma onto the Sun in a highly condensed manner. When the plasma-focusing system becomes inactive, below the required threshold conditions, the Sun reverts to a type of cosmic default level with

70% less energy being radiated, and higher rates of solar cosmic-ray flux being experienced.

At the present rate of plasma diminishment being experienced, the solar activity phase-shift threshold to the next Ice Age period may be crossed in 30 years, or in the 2050s, most likely. With the primer-fields system gone inactive by then, the climate on Earth will get 40 times colder than the Little Ice Age in the 1600s had been. Ice core evidence promises that. Without the needed preparations for human living in such an environment, 99% of humanity would die of starvation, both by the cold, and by CO_2 depletion that diminishes agriculture, as more CO_2 becomes dissolved into the sea.

With the 'Primer Fields' being critical for our very existence, the exploration of them is likewise critical.

In the Little Ice Age, between 10% and up to 30% of the populations in Europe had perished by starvation. The last Big Ice Age was evidently vastly harsher. Only 1-10 million people emerged from it alive. That's all we had after 2 million years of development. We want to do far better this time around; and we can, with large-scale technological infrastructures for our food supply. But will we create them? Will we get the job done in the 30 years that we still have left before the Ice Age starts anew? Will we even consider it? And how certain are we that the phase shift to the next glaciation period will begin, as the evidence suggests, in the 2050s? We have no slack on this front. Should we fail us on this absolute front, we would be committing suicide.

Numerous fields of evidence tell us that the next Ice Age is near. That's where the truth begins. Most of the evidence was discovered in the 1990s and thereafter. Some evidence is measured in ice cores; some is measured in space, by satellites. Some measurements are also made on the ground in terms of measurements of the Earth's magnetic-pole drift observed in northern Canada. All of this is seen combined with high-energy physics experiments at a leading national laboratory, and is also explored in the small in static experiments.

So, what will the answer be? Will we move with the evidence? Or will we lay ourselves down to die by default?

It takes an independent researcher to brake the taboos that have kept mainstream cosmology imprisoned, increasingly, during the past century, even while what is regarded as taboo is known to be wrong.

The Illustrated Science series is intended to open the scene beyond the threshold of accepted taboos, to where the actual physical evidence speaks for itself.

The scope of the existential challenge that the Ice Age brings with it, takes astrophysics out of the academic domain and places it into the foreground as one of the most-critical issues of our time. The big Climate Change events that have already worldwide effects are mere fringe effects in the flow of the ever-changing cosmic dynamics. The big effect, when the Ice Age begins anew, promises to be caused by a dimmer and colder Sun. The loss of 70% of the Sun's radiated energy defines our climate future that begins in the near term.

Sure, we can live with all that by creating new platforms for agriculture that are able to operate under Ice Age conditions. But will we do it? The task is enormous. Or will we fail ourselves on this front? We have no reason to allow us to fail. We have the materials and energy resources on hand to accomplish everything that is required for us to continue to live in an Ice Age World. But will we do it? The big question that never goes away, therefore, is; will we develop our inner resources as human beings sufficiently to get the job done, and to get it done in time? Or will we do nothing, ignore the challenge, and condemn our children and one-another to an agonizing death by starvation? That's the choice.

Towards meeting the inner challenge, I have created the epic series of novels, The Lodging for the Rose. And further, towards meeting the science challenge, I have produced numerous research books and several dozen exploration videos that the Illustrated Science series is modeled after. The work is the result of a quarter century of research, for which numerous elements of evidence in related fields came to light during the timeframe of my research.

It is my hope that the work that went into all of these projects will help in some degree - for humanity that we are all a part of - to write itself a ticket to have a future.

10

High-resolution color images, of the images in this book, can be obtained at www.iceagetheatre.ca

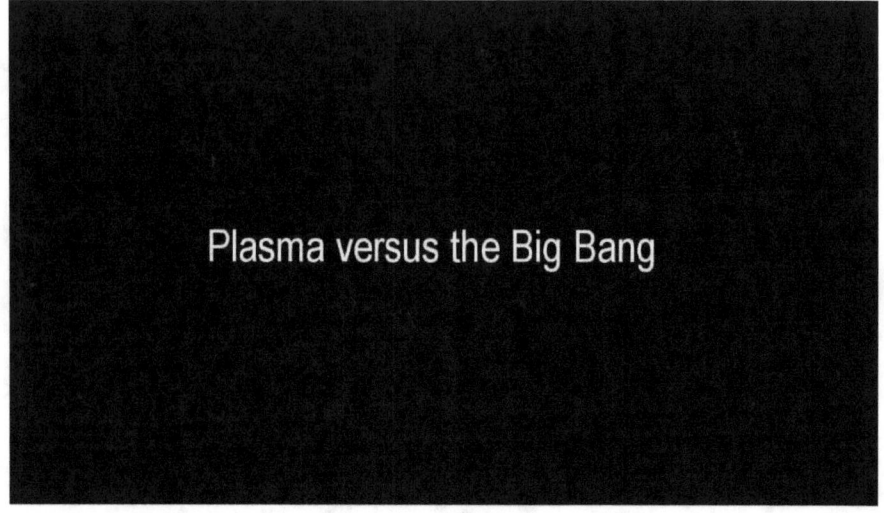

Our Electric Cold Fusion Sun (Part 2) Plasma vs. the Big Bang

The Big Bang creation is more a myth than a theory

The Big Bang creation is more a myth than a theory. It may be termed a science-dream on the order of fairy tales.
The dream begins with nothing. Before the Big Bang happened, there was nothing. The universe did not exist. Reality was an inconceivable dark void.

The entire universe exploded into being

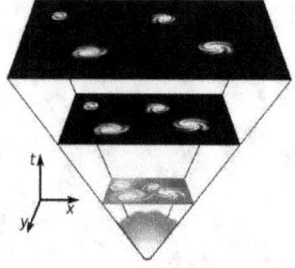

The Big Bang creation myth refuted
by the electric solar fusion model

Suddenly in this infinite nothing, 13.8 billion years ago, the entire
universe exploded into being.

All mater, all energy

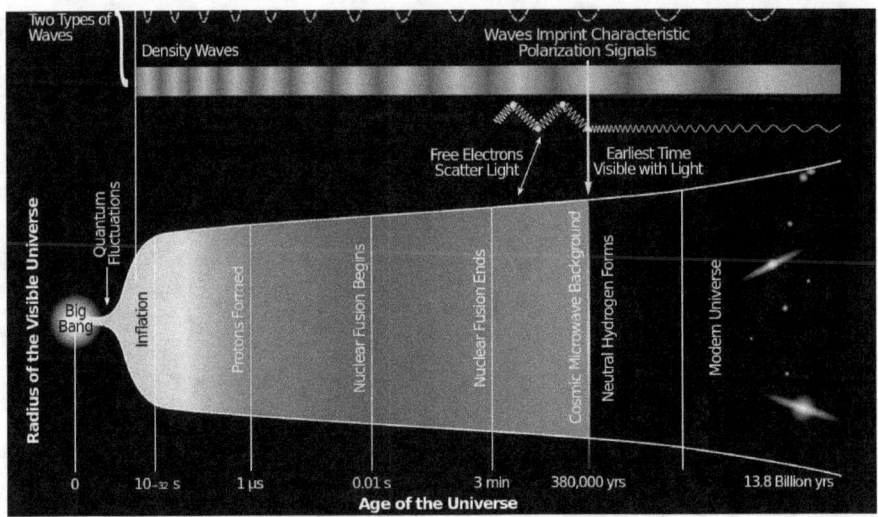

All mater, all energy, all that there is and ever will be, is deemed to have emerged in the space of three minutes out of the 'whirlpool' of the giant explosion that expanded into all directions from its central point as a primordial cloud of dust that is still expanding outward with ever-increasing speed.

The dust is deemed to have condensed

NASA-Hubble Ultra Deep Field
2014-2014 0603

The dust is deemed to have condensed into galaxies and stars and planets. From within the dust, the hydrogen atoms are said to have grouped together into giant spheres of gas that became stars.

The tale is full of holes

A deemed star-forming region in the Large Magellanic Cloud. NASA/ESA image

It is deemed that each star, by its accretion of weight, became so hot under the resulting intense gravitational compression that the hydrogen atoms in its core began to fuse into larger helium atoms. The building of larger atoms is deemed to create energy, from which the stars derive their light. Thus each gas star became a Sun, powered from within by its own substance.

All this comes together as a nice tale that almost makes sense. But the tale is full of holes, holes created by paradoxes. Numerous impossible paradoxes, supported by physical evidence, solidly refute the impossible theory.

The Earth itself, is a paradox under the Big Bang theory

The Earth is our witness that atom-synthesizing electric nuclear fusion is continuously occurring

NASA

The existence of the Earth itself, is a paradox under the Big Bang theory. It shouldn't be what it is. Its composition disproves the theory that the universe was created in a primordial explosion 13.8 billion years ago.

Light disproves the explosion theory of the Big Bang

2MASS LSS chart-NEW Nasa

The center of the Milky Way, at the center of the Big Bang explosion of the universe

Even the nature of light disproves the explosion theory of the Big Bang, in which the formed dust condensed it into galaxies that are deemed to have expanded across the cosmos, and are still expanding, and racing away from us into all directions with accelerating motion.

The red-shift of light from distant galaxies

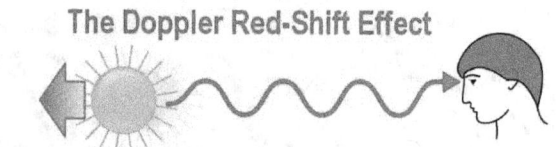

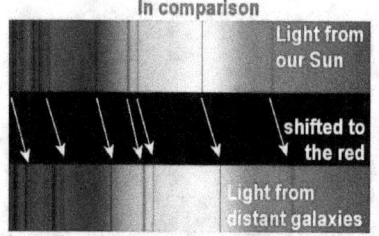

As proof for the accelerating motion, a theory has been invented that interprets the spectral red-shift in light from distant galaxies as being caused by the source of the light racing away from us, the observer.

The red-shift of light from distant galaxies is said to prove the Big Bang theory of the accelerating expansion of the universe. Light is said to be stretched out as the most distant galaxies are deemed to be racing away faster and faster, the further they are located from us. That's the tale that is being told. It is almost believable.

Red shift is the result of energy depletion

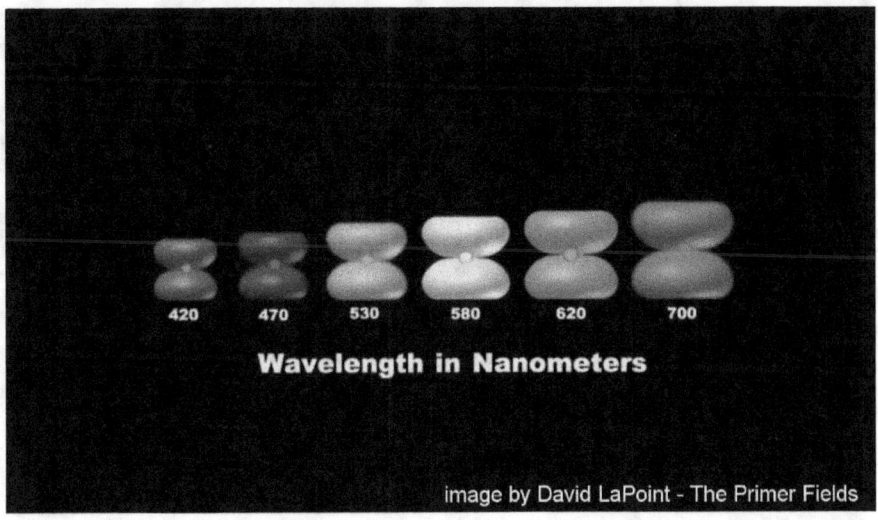

420   470   530   580   620   700

Wavelength in Nanometers

image by David LaPoint - The Primer Fields

In the real world the theory breaks down when the measured red shift is more rationally perceived as the result of energy depletion of the photons over long distances in their encountering plasma and atomic gases in intergalactic space. The energy depletion causes the entire electromagnetic spectrum to shift towards a lower energy state, which means a shift towards the red. The photons of light are said to be packages of energy. The more energy they contain, the tighter the packages are packed. When energy is dissipated, the packages become larger. The violet expands into blue, the blue into green, the green into yellow. The entire spectrum shifts towards the red. This simple reality, disproves the Big Bang creation theory.

A gigantic piece of evidence that disproves the Big Bang theory

The Earth from 98,000 miles during the Apollo 11 translunar injection on July 16, 1969

NASA

We also have a gigantic piece of evidence that disproves the Big Bang theory even more, and proves at the same time that every sun is a plasma-powered electric nuclear-fusion engine that synthesizes all atomic elements that are known to exist in the universe. This item of evidence is the Earth itself. The proof for this statement was delivered only recently in a massive scientific effort to determine the age of the Earth.

The Earth is proof that synthesizing atomic fusion

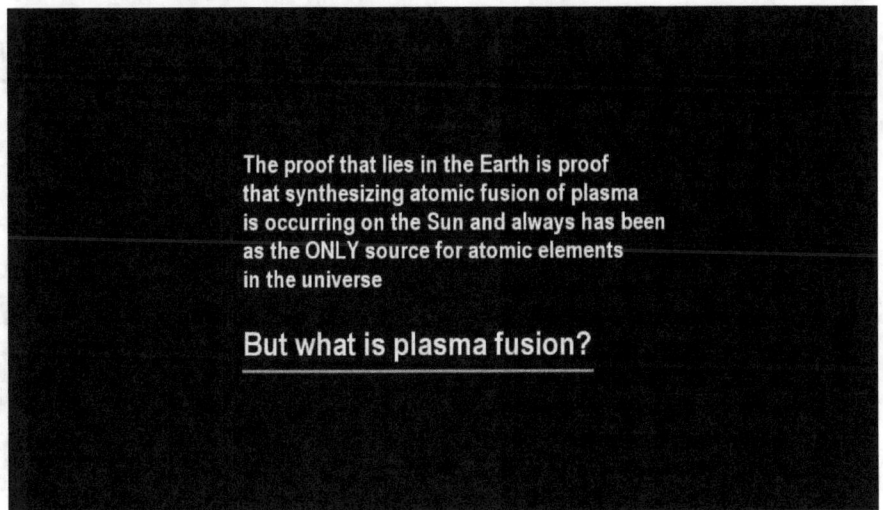

The proof that lies in the Earth is proof that synthesizing atomic fusion of plasma is occurring on the Sun and always has been as the ONLY source for atomic elements. But what is plasma fusion?

Plasma is a sea of electrons and protons

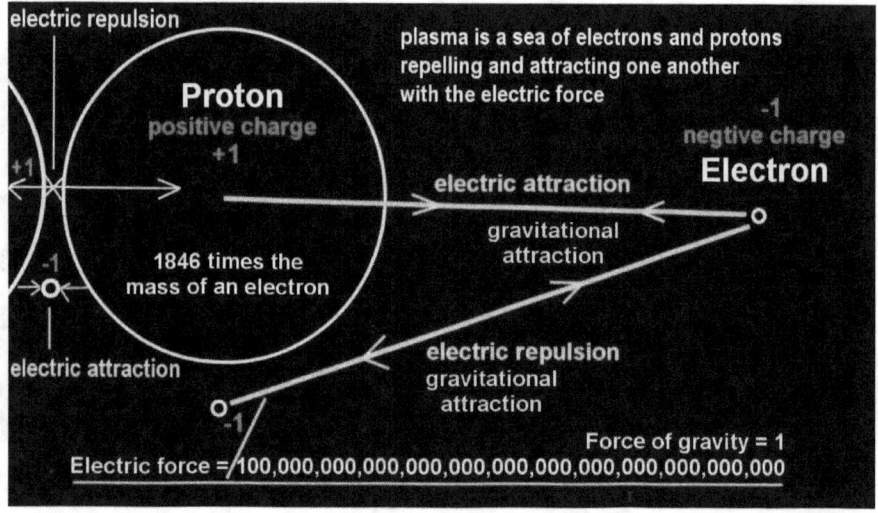

Plasma is a sea of electrons and protons repelling and attracting one another with the electric force.

*Plasma in space

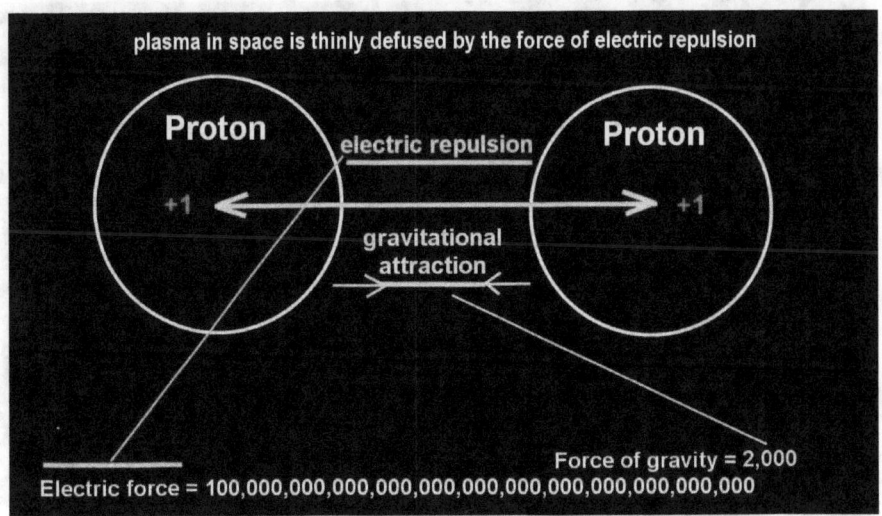

Plasma in space is thinly defused by the force of electric repulsion.

On contact, two protons snap together

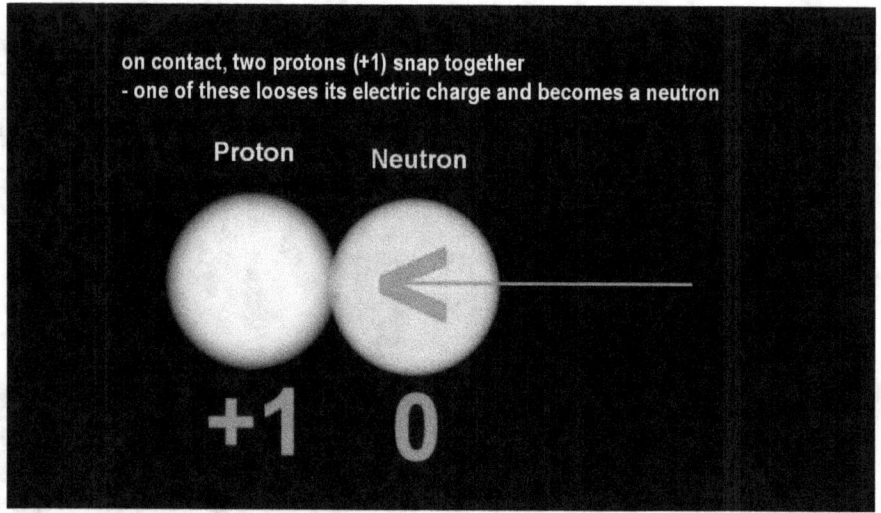

On contact, two protons snap together. 'Thereby, one of the protons looses its electric charge and becomes a neutron.,

On contact, the electron is forced to rebound

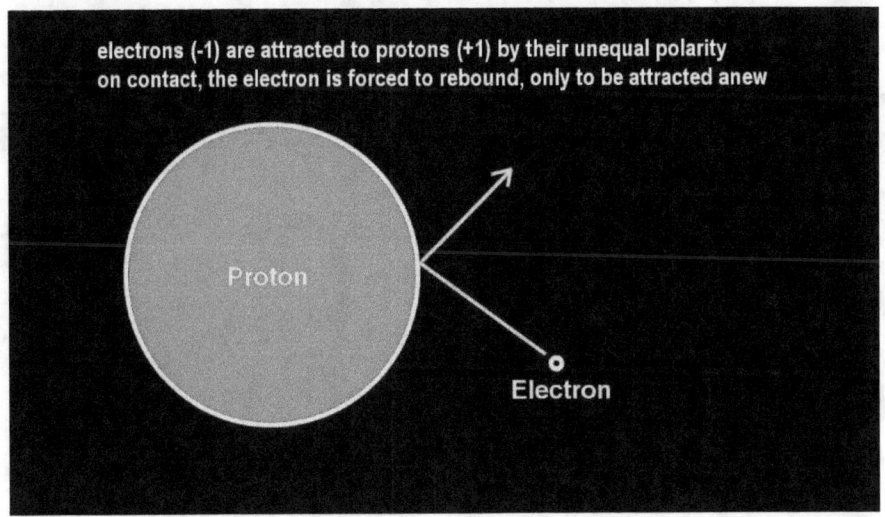

electrons (-1) are attracted to protons (+1) by their unequal polarity
on contact, the electron is forced to rebound, only to be attracted anew

Electrons (-1) are attracted to protons (+1) by their unequal polarity. On contact, the electron is forced to rebound, only to be attracted anew.

Atoms are formed by the dynamic 'dance'

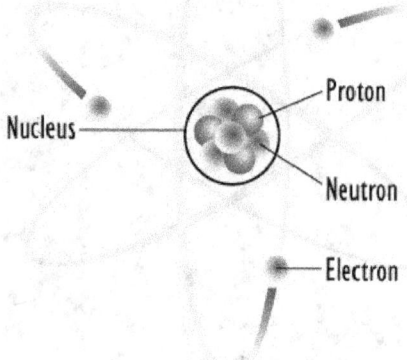

**atoms are formed by the dynamic 'dance'
of electrons being attracted and forced to rebound**

Nucleus

Proton

Neutron

Electron

wikipedia (image)

Atoms are formed by the dynamic 'dance' of electrons being attracted and forced to rebound.

Atoms are electrically neutral plasma structures

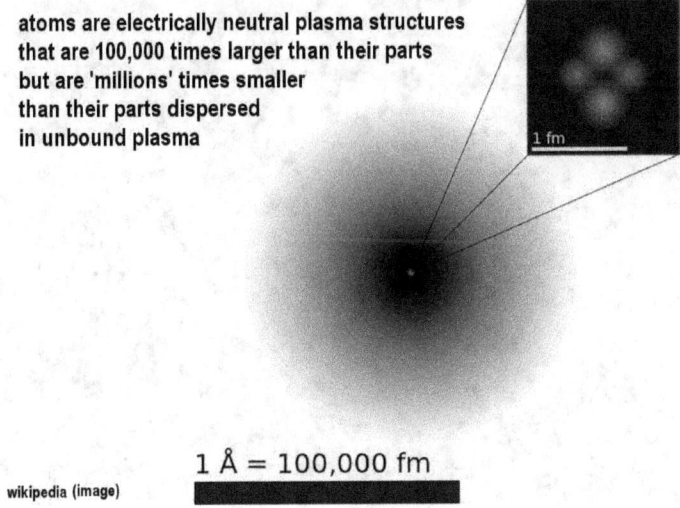

atoms are electrically neutral plasma structures
that are 100,000 times larger than their parts
but are 'millions' times smaller
than their parts dispersed
in unbound plasma

1 fm

$1 \text{ Å} = 100,000 \text{ fm}$

Atoms are electrically neutral plasma structures that are 100,000 times larger than their parts, but are 'millions' times smaller than their parts dispersed in unbound plasma.

Atom-forming fusion increases mass density

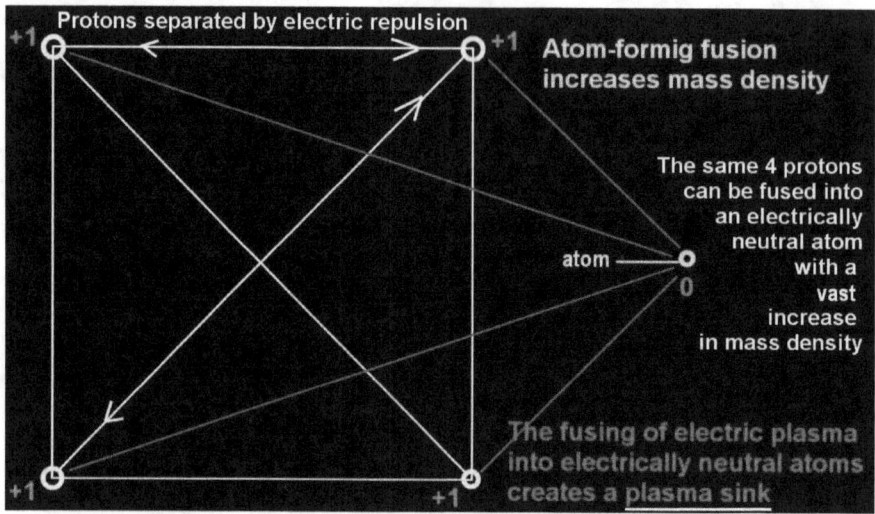

Protons separated by electric repulsion

Atom-formig fusion increases mass density

The same 4 protons can be fused into an electrically neutral atom with a vast increase in mass density

atom — 0

The fusing of electric plasma into electrically neutral atoms creates a plasma sink

Atom-forming fusion increases mass density, while the forming of electrically neutral atoms creates a sink in the electrically charged landscape.

Electric nuclear fusion happens naturally

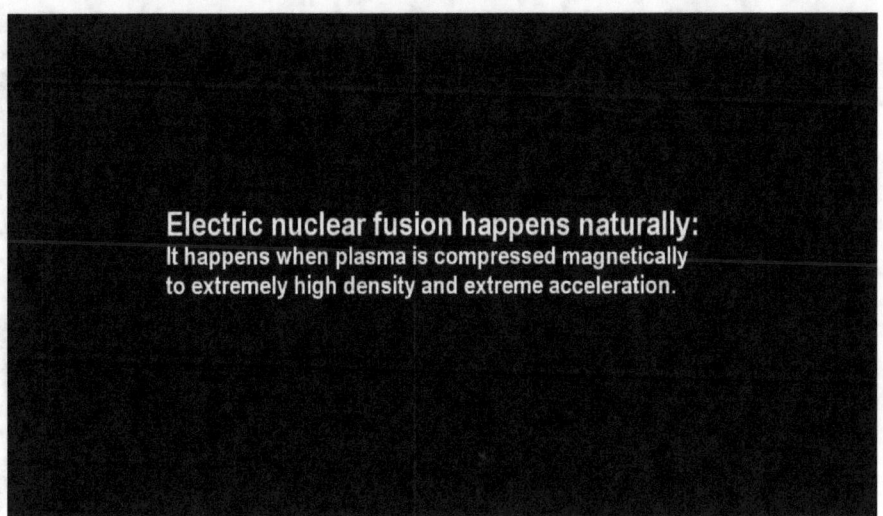

Electric nuclear fusion happens naturally: It happens when plasma is compressed magnetically to extremely high density and extreme acceleration.

Experiments at the Los Alamos National Laboratory

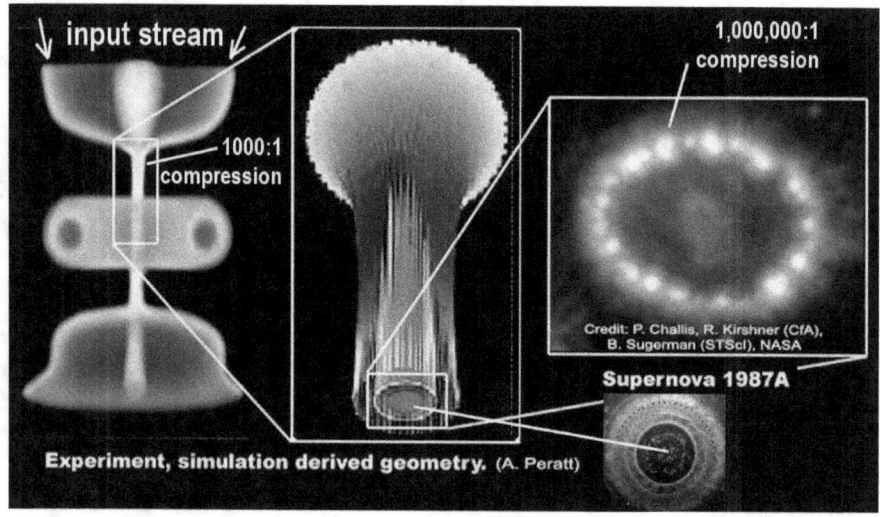

Experiments at the Los Alamos National Laboratory illustrate that a million to one plasma compression is natural in high-energy, magnetically focused plasma flow dynamics.

Plasma compression may be a billion-fold

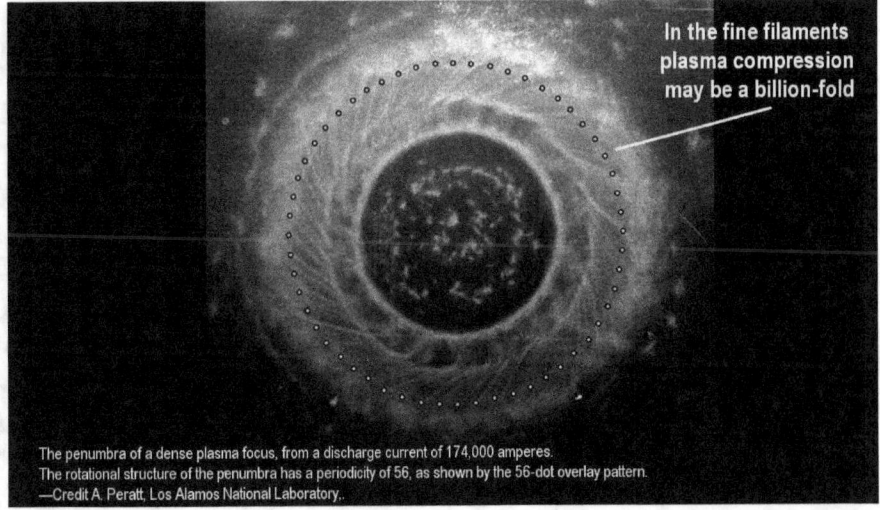

In the fine filaments plasma compression may be a billion-fold

The penumbra of a dense plasma focus, from a discharge current of 174,000 amperes.
The rotational structure of the penumbra has a periodicity of 56, as shown by the 56-dot overlay pattern.
—Credit A. Peratt, Los Alamos National Laboratory.

In the fine filaments, plasma compression may be a billion-fold.

Very large cosmic 'primer fields'

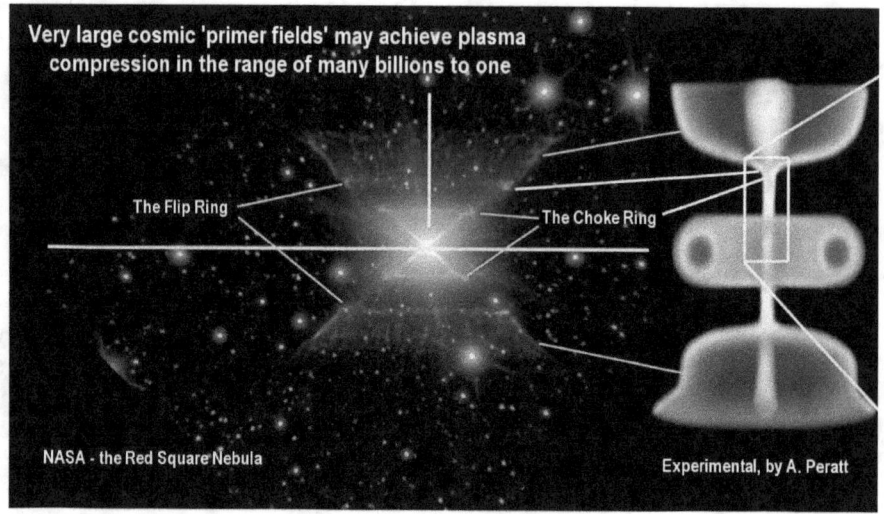

Very large cosmic 'primer fields' may achieve plasma compression in the range of many billions to one.

Plasma compression may exceed trillions to 1

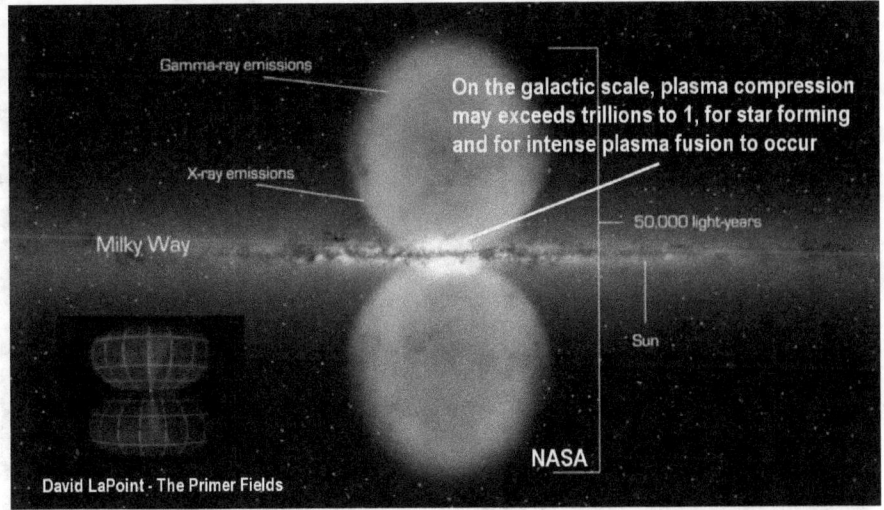

On the galactic scale, plasma compression may exceed trillions to 1, for star forming and for intense plasma fusion to occur.

The Earth is our witness that atom-synthesizing electric nuclear fusion is continuously occurring.

Large atomic elements decay over time

Radioactive decay chains

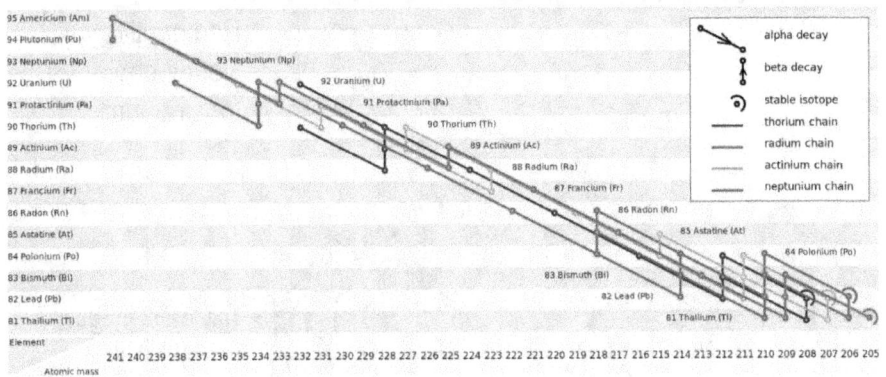

It had been discovered previously that large atomic elements, such as uranium, decay over time in a chain of mutation that ends in lead. Lead is the heaviest stable element that exists. It has been further discovered that the rate of decay is knowable for the different decaying elements.

The ratio of lead in uranium-containing rocks

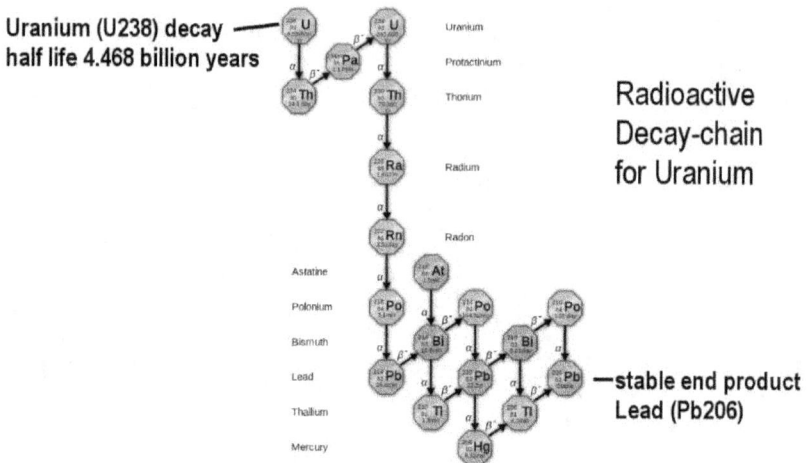

Uranium, for example is known to have a half-life of 4.5 billion years. This means that over the span of 4.5 billion years, half of the volume of uranium that existed in the beginning, has become transmuted into lead.

It has been reasoned that the measurable ratio of lead in uranium-containing rocks, can be used to accurately determine the age of the Earth. On this basis the age of the Earth, and of the solar system as a whole, has become known to be in the range of 4.54 billion years.

It has been discovered that the lead-to-uranium ratio is always the same, wherever uranium is found, including in cases were uranium is found in the remains of asteroids.

The measured ratio disproves the Big Bang theory

Milky Way look-alike
NGC 6744

The measured ratio disproves the Big Bang theory. The ratio proves that all the atomic elements on Earth and in the solar system nearby, were NEWLY created, at the time the Earth and the solar system was formed, which occurred likely near the center of our galaxy where the plasma pressure is strong and the fusion process on the surface of the Sun is immensely productive, so much so, that even the largest atoms were synthesized in substantial quantities, such as uranium.

The dating of the Earth, with the atomic clock

The Earth is formed

At time zero

Uranium

Today 4.5 billion years later

Lead	Uranium

The scientific dating of the Earth, that has been timed with the atomic clock of the universe, proves that all the atomic elements that the Earth is made of, did not originate as materials that were formed in the Big Bang, more than 13 billion years ago. The radioactive decay would have produced a much greater ratio of led in uranium-containing rocks. The measured ratio proves that all the atoms for the planets in the solar system were synthesized NEW at the time of the forming of the solar system.

This means that the synthesis could only have occurred on the surface of the Sun, in the time of its initial intense state. While atomic-fusion synthesis still occurs on the surface of the Sun today, it does so with lesser intensity.

The dating of the Earth, with the atomic clock of the universe, doesn't match the Big Bang creation theory by a long way. It thereby disproves it. It leaves the external-fusion Sun standing as the only possible contender for the synthesis of the atomic elements in the solar system.

So it is that the scenario that solves the paradox, is a case of actual historic evidence.

Gravitational accretion of cosmic dust from the Big Bang

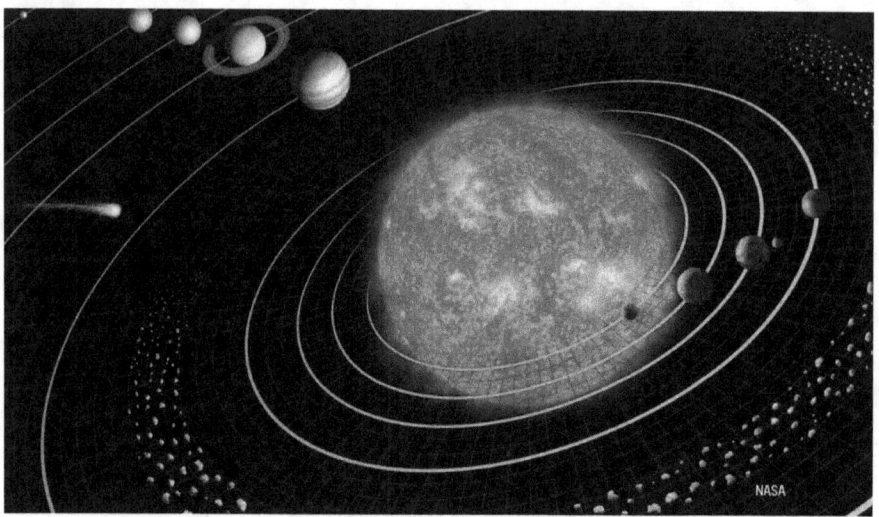

The fabled theory that all the stars and planets were drawn together by the gravitational accretion of cosmic dust from the Big Bang explosion, doesn't make any sense anyway. The theory is a paradox in itself. No principle exists that would single out the hydrogen atoms from the cosmic dust to gravitationally form a star. Hydrogen atoms have the least gravitational attraction of all the atoms in the universe. Even helium asserts a 4 times greater gravitational attraction than hydrogen. Shouldn't all the stars be made up of helium then? And what about the really heavy elements with huge gravitational attraction, such as iron, lead, or even uranium? Shouldn't they form the core of the stars? The entire hydrogen-sun theory is a paradox. It doesn't have a basis to stand on that makes any sense.

The hydrogen-sun theory

The hydrogen-sun theory is so evidently false that it is surprising that it is still maintained in the empty box of the gravity-only universe represented by the Big Bang dream.

Look at the volume of hydrogen that is needed

Just look at the volume of hydrogen that is needed to fill up the Sun, and also the four gas planets that are almost entirely made up of hydrogen. This enormously huge volume of hydrogen does not exist, according to the cosmic abundance ratio that has been measured on the Sun and throughout the universe.

According to the false theory

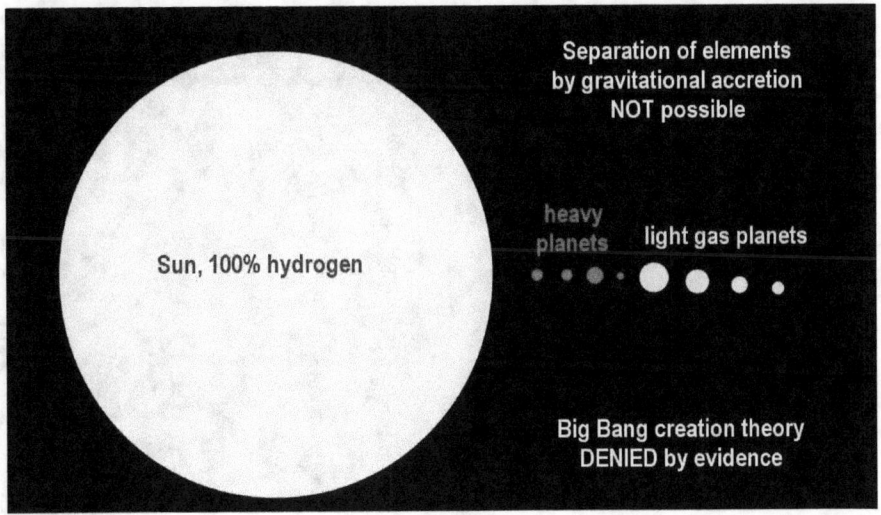

According to the false theory the Sun started as a hydrogen star, which has converted hydrogen into helium for more than 4 and a half billion years, towards its present ratio.

Where did the huge volume of hydrogen come from?

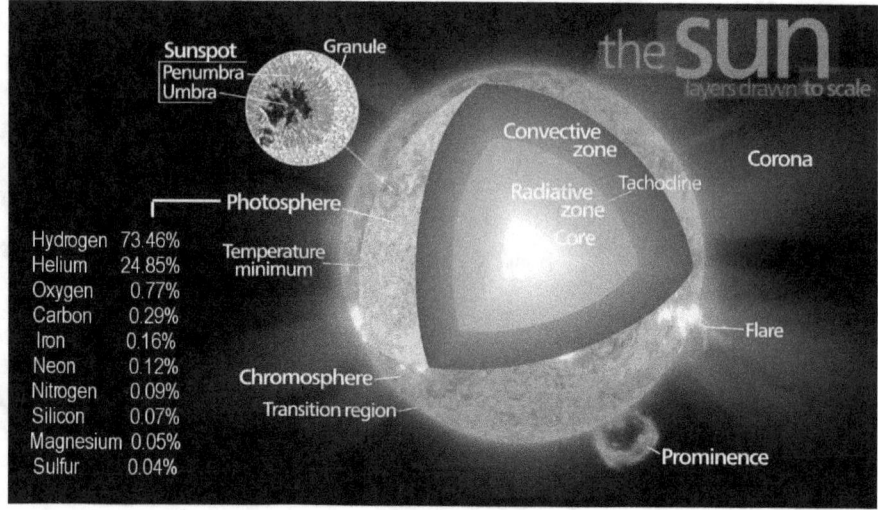

Where did the huge volume of hydrogen come from that the Sun, supposedly, has converted into helium over those billions of years of its operation at an estimated hydrogen conversion rate of 620 million metric tonnes per second?

Out of the range of the cosmic abundance ratio

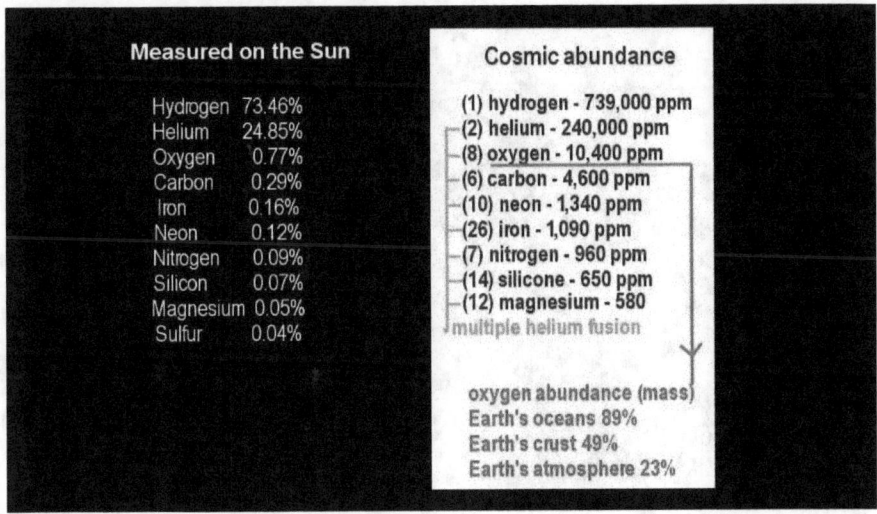

The resulting dynamic consideration places the hydrogen ratio that would have had to exist at the birth of the Sun, way out of the range of the measured cosmic abundance ratio. This impossible paradox can only be resolved on the basis of recognizing that the Sun is not, and never has been, a sphere of hydrogen and helium gas, but exists as a sphere of plasma. With this fundamental correction made in the theory, the gas ratio in the solar system closely reflects the measured cosmic abundance ratio.

The case of comparison also illustrates what it is that we actually see, when we look at the Sun through the umbra of the sunspots.

*We see a Sun that is dark inside

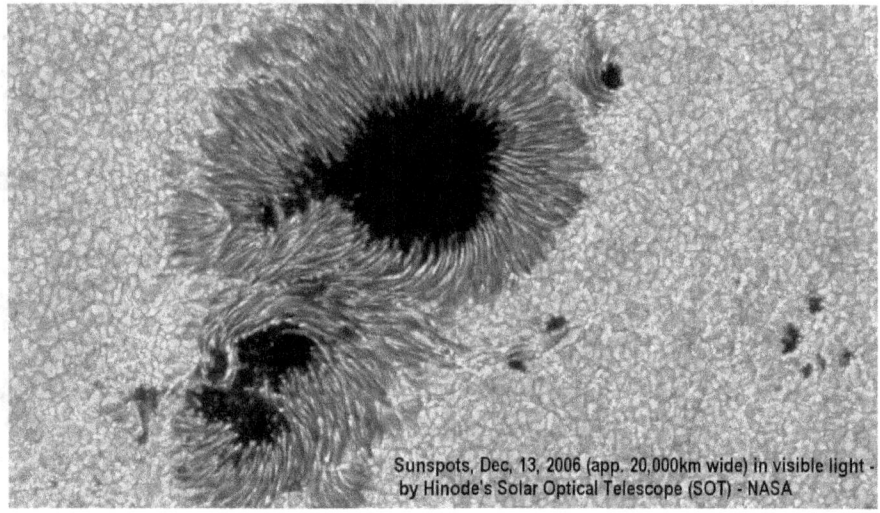

Sunspots, Dec. 13, 2006 (app. 20,000km wide) in visible light - by Hinode's Solar Optical Telescope (SOT) - NASA

When we look at the Sun, and look through the umbra of the sunspots, we see a Sun that is dark inside. We can see plainly that internal nuclear fusion isn't happening in the Sun. The Sun would be brighter inside if nuclear fusion would be happening internally, as the solar power source.

A Sun that is a sphere of thinly dispersed plasma

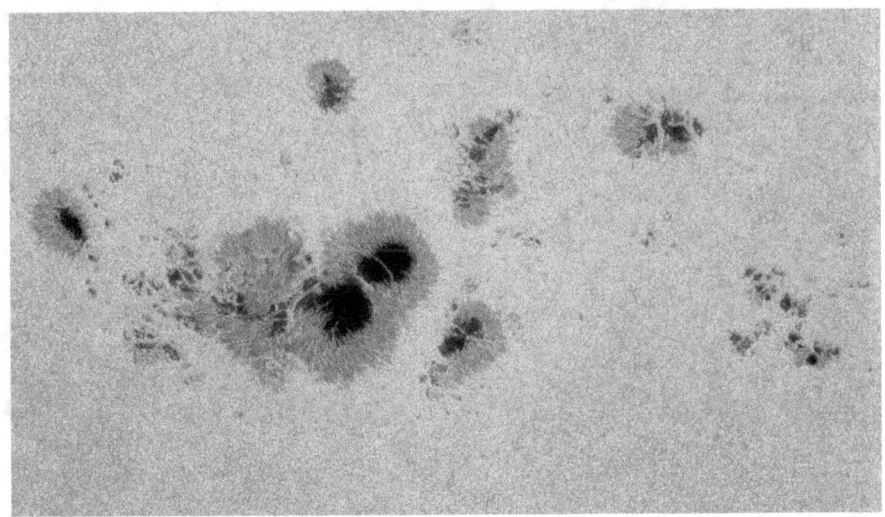

We see instead that everything happens on the surface of the Sun. This means that what we see in the Sun, is exactly what we should see for a Sun that is a sphere of thinly dispersed plasma with electric fusion occurring on its surface. We see that nothing exists there to be seen, below the surface. Plasma particles, which are 100,000 times smaller than an atom is, are invisible. They emit no light. Plasma is dark. It is a type of black hole. That's what we see. Of course, by looking at the Sun, we can also see that the Sun is intensely energetic on its surface.

From where does the Sun derive its energy

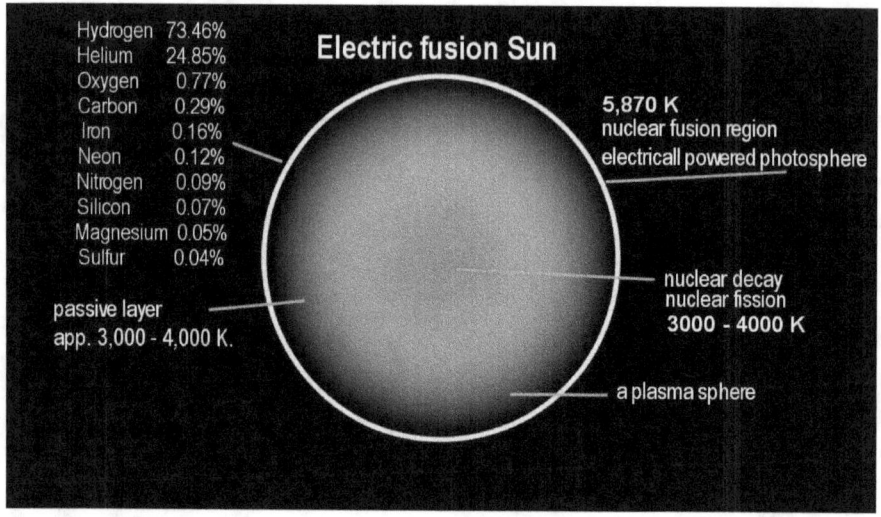

From where then, does the Sun derive its energy and its substance for synthesizing fusion? Where would the energy come from, if it didn't come from external plasma flowing into the Sun?

From plasma our world was formed!

The Sun at
a plasma node

David LaPoint - The Primer Fields

The answer is simple. The Sun's energy and its fusion-input are both exclusively derived from plasma.
From plasma our world was formed!
But where does the plasma come from? Is it the product of the Big Bang? Evidently it isn't since it is well demonstrated that the Big Bang theory is a science fairy tale. So, where did the plasma come from, and still does come from?

The radiometric dating of the Earth

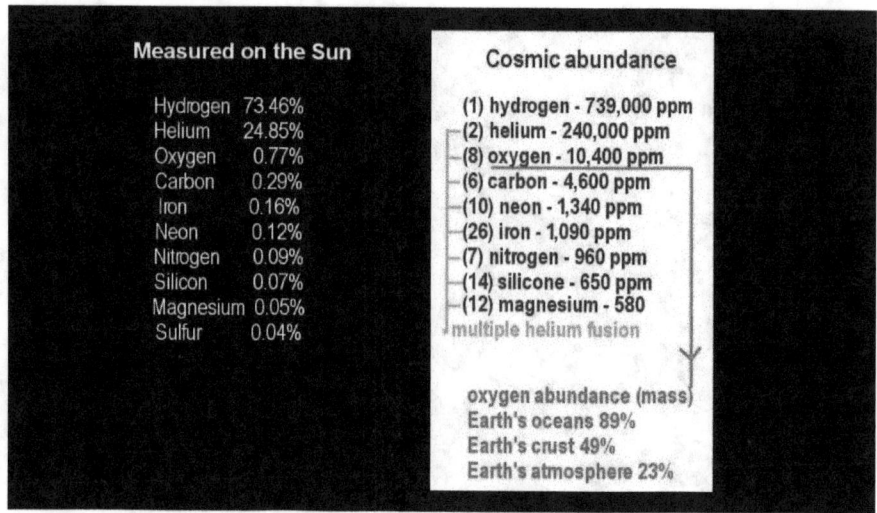

The radiometric dating of the Earth with the atomic clock of the universe, proves the solar synthesized creation of the Earth to be real, as a construct of fused plasma. We are looking at enormous volumes of plasma flowing into the Sun for this gigantic creative process to be possible.

Plasma gets 'consumed' by the solar electric-fusion process

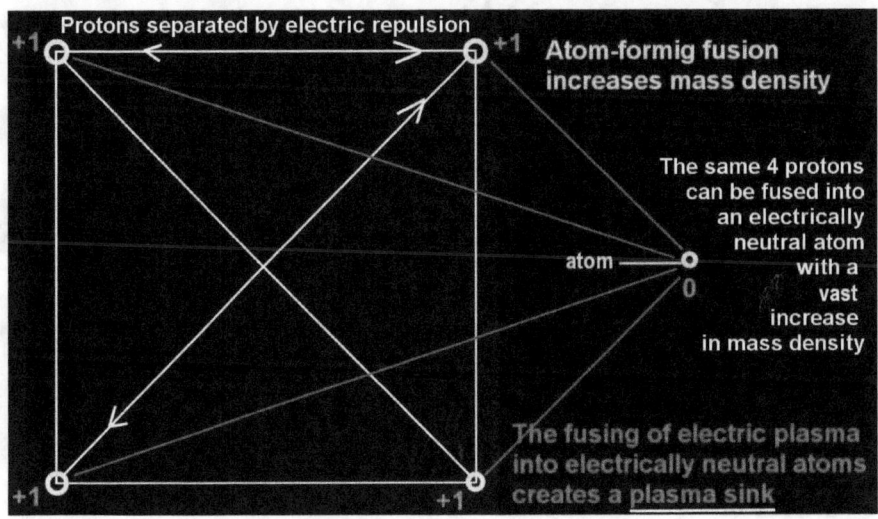

It is interesting to note that free plasma gets 'consumed' by the solar electric-fusion process, as plasma is converted into atoms. The consumption of plasma by the fusion process, creates a plasma sink that gets plasma moving. Plasma has no energy in itself, but becomes energetic when it is set in motion. And for plasma to move, we need a source and a sink that it flows into. The cold fusion process is this sink. But what is the source? We have monumental evidence on hand for the sink process, and almost no evidence that there is a source.

Evidently it is not enough for the Sun to have streams of plasma to flow to it, to energize it, because every physical flow process needs a source and a sink for it to function. The fusing of protons into electrically neutral atoms, renders the Sun a plasma sink.

If plasma would merely flow into the Sun

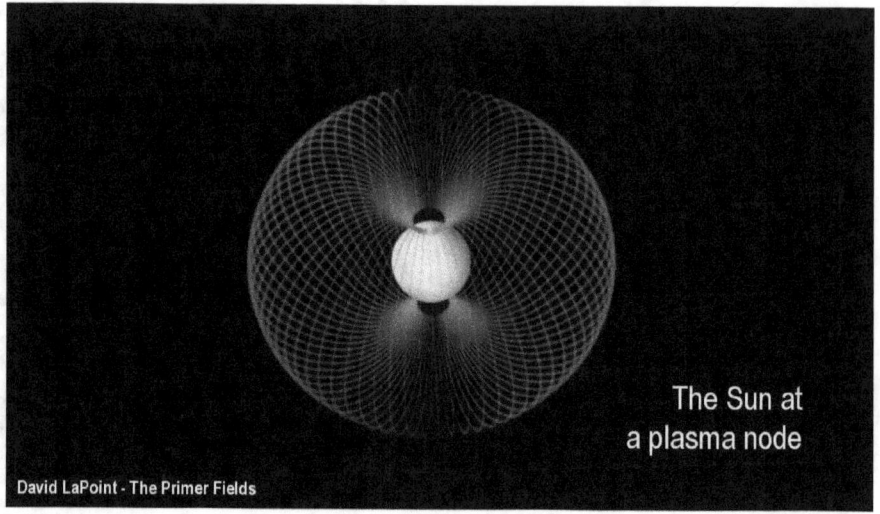

The Sun at
a plasma node

David LaPoint - The Primer Fields

If plasma would merely flow into the Sun to energize it, it would simply pile up there, and nothing would happen. Plasma needs to flow to be intensely energetic, just like water needs to flow for hydro-electric generating systems to work.

For hydro-electric generating to work

Grand Coulee Dam

For hydro-electric generating to work, water needs to flow from a high point, the source, to a low point.

The energy of the water

water turbine at the
Grand Coulee Dam

The energy of the water, flowing from a high point, pulled by the force of gravity, delivers the energy that turns the turbine.
The energy that is generated on the Sun must have a similar potential. It must have a source and a sink, for an energetic flow to happen. On the Sun, the conditions are met. The nuclear fusion process energizes the dynamics that also create energy in the form of light and heat.

The synthesizing fusion on the surface of the Sun

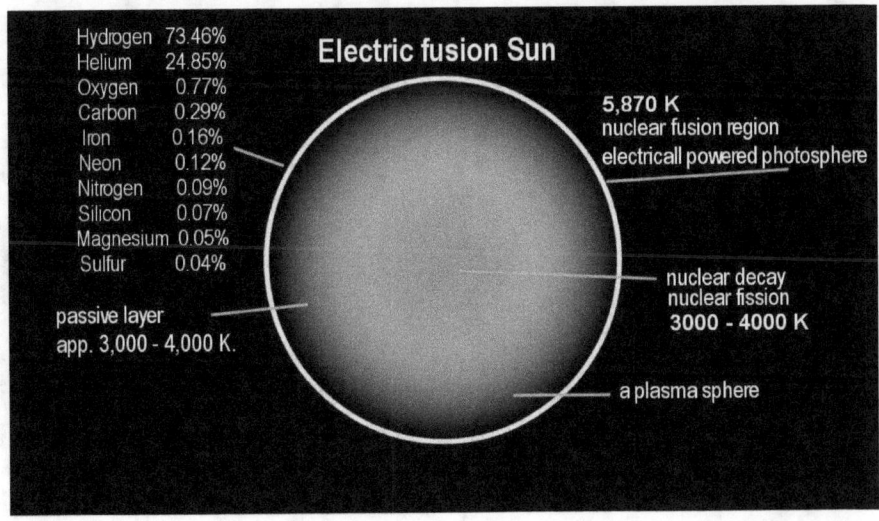

The synthesizing fusion on the surface of the Sun literally eats up plasma, packs it together, and spits it out as electrically neutral atoms that flow away with the wind. Without this process, of the Sun eating plasma, nothing would happen. Nothing would flow. The process may be termed 'cold' fusion, as the process is not initiated by intense thermal agitation in the range of millions of degrees, but functions by cold electro-magnetic plasma acceleration and plasma pressure that also produce a modest amount of thermal and light energy along the way at a temperature of 5,505 degrees Celsius.

Plasma-fusion maybe the sink that activates the source

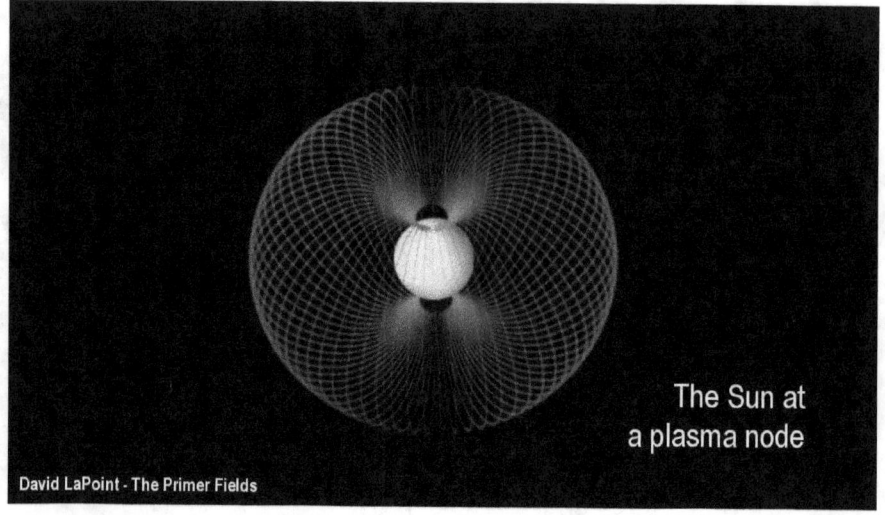

The Sun at
a plasma node

David LaPoint - The Primer Fields

Electric plasma-fusion maybe the sink that activates the source.
The Sun and its fusion process is easily identified by its functioning
as a consuming sink. In comparison the source is far-more vague,
and difficult to conceptionalize. Nevertheless, this too, is of critical
importance to us living on this planet.
Even in hydro-electric generating system, the source of the water
that delivers the energy is often miles distant and is rarely
perceived as related to the sink system.

The faucet
as a sink

wikipedia

When the plasma-flow into a sink activates the source, no matter how far away the source may be located, just as it would in a water-supply system, then a faint picture for a possible concept for a plasma source comes to light. For the water-supply system, the source may be a lake far distant in the mountains.

The American theoretical physicist David Bohm

The Implicate Order and Explicate Order

David Bohm
1917 - 1992

For the plasma supply system, the source may simply be the vast expanse of space, that, as it is defined by the American theoretical physicist David Bohm, is not really empty space at all. He introduced the concepts of Implicate Order and Explicate Order, which appears to be exotic theory, but according to evidence may be most fundamental to everything. David Bohm stated that "Space is not empty. It is full, a plenum as opposed to a vacuum, and is the ground for the existence of everything, including ourselves. The universe is not separate from this cosmic sea of energy." The explicate is then merely a specific expression of enfolded implications that leave on the surface but ripples of countless waves of energy that coming together take on a specific form. Since each wave carries an amount of energy, it has been theorized that a cubic inch of space may contain more energy than is found in all the galaxies of the universe.

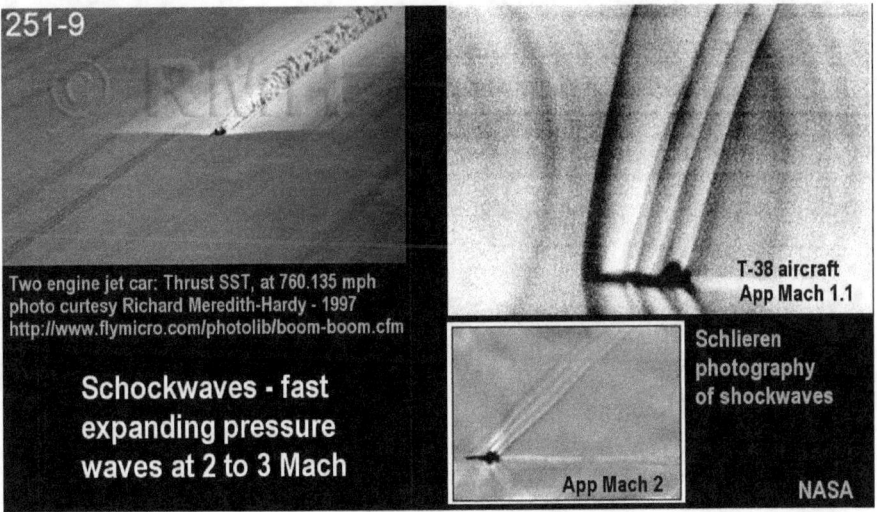

251-9

Two engine jet car: Thrust SST, at 760.135 mph
photo curtesy Richard Meredith-Hardy - 1997
http://www.flymicro.com/photolib/boom-boom.cfm

Schockwaves - fast expanding pressure waves at 2 to 3 Mach

T-38 aircraft
App Mach 1.1

Schlieren photography of shockwaves

App Mach 2

NASA

That this concept may not be far off the mark is evident in the speed limit of light as it propagates through space. The speed of sound in the air is limited by the fluid dynamics of the air. The speed of light may be similarly determined by a type of 'fluid' dynamics that governs the vast energy background in what is deemed empty space.

In the sea of latent energy that is cosmic space

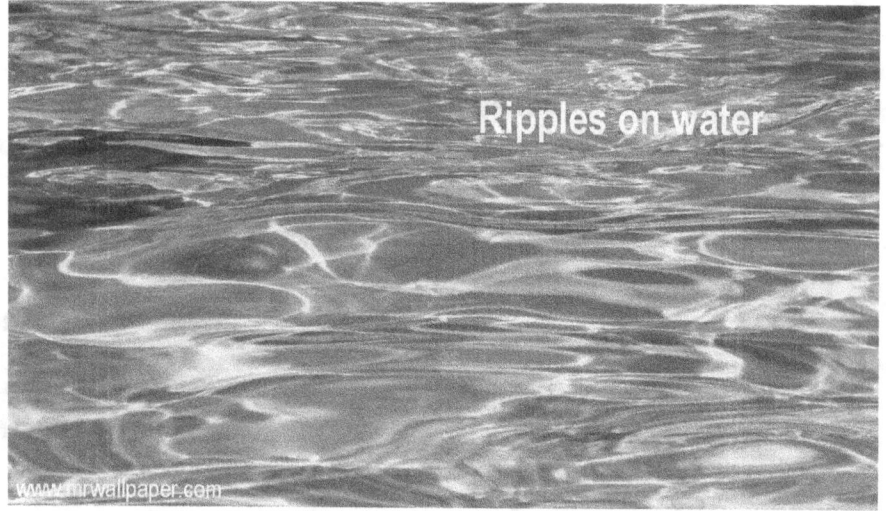

In the sea of latent energy that is cosmic space, the ripples that form explicitly on the 'surface' may form the quarks that in turn form the electrons and protons that make up the plasma in the universe.

Quarks cannot be divided

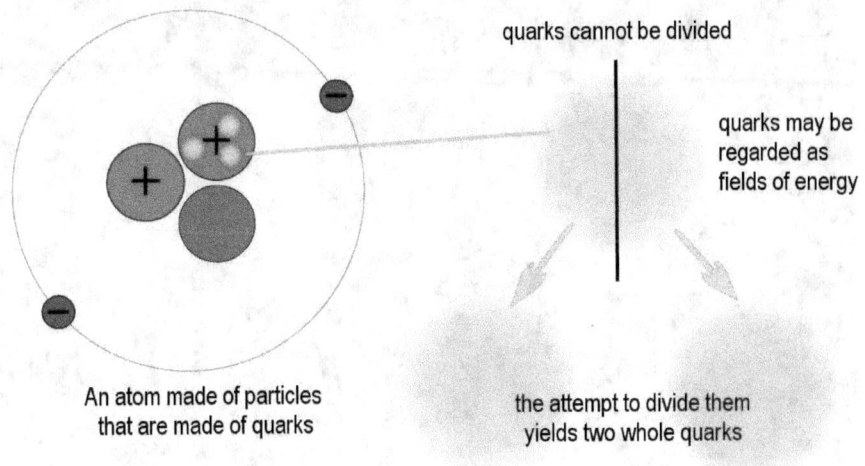

quarks cannot be divided

quarks may be regarded as fields of energy

An atom made of particles that are made of quarks

the attempt to divide them yields two whole quarks

It is interesting to note that quarks cannot be divided. An attempt to break them apart, yields not two halves, but two whole quarks.

David Bohm,

As exotic as this may sound, the originator of the Implicate Order and Explicate Order, David Bohm, who lived between 1917 and 1992, may have solved this puzzle for us, and also the puzzle of the origin of plasma. David Bohm is considered to be one of the most significant theoretical physicists of the 20th century, whom Einstein had referred to, as his successor.

We see two very-long climate cycles expressed

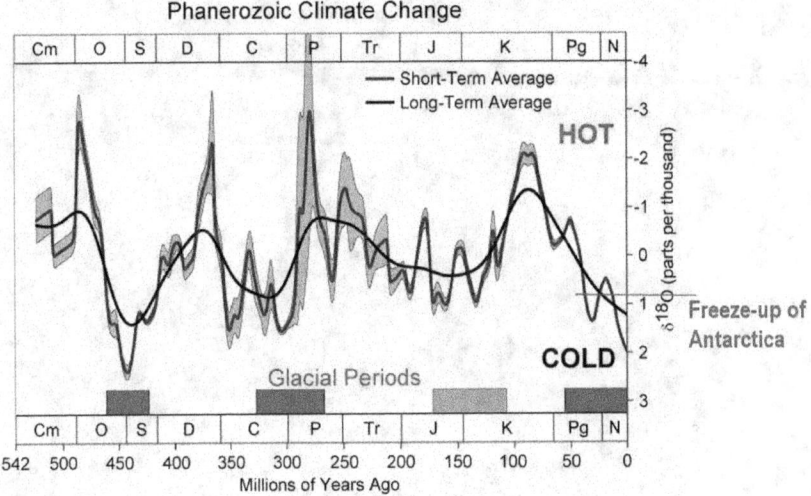

Phanerozoic Climate Change

Exotic as the theory may seem, we have some amazing physical evidence for it, which is expressed in the way our climate on Earth has been modulated in long cycles over the last 500 million years. We see two very-long climate cycles expressed here, that are overlaid over each other. We see a 150-million-years cycle, and a 31-million-years cycle expressed. These are huge cycles with enormous effects. Note, where on this gigantic scale the freeze-up of Antarctica is located, and where the current stage of the world is located, that puts us at the lowest and coldest point in 440 million years.

The very long cycles can be seen as evidence

The Capodimonte Deep Field - of 35,000 galaxies

NASA

The very long cycles can be seen as evidence that the plasma that powers our Sun may have its origin primarily in intergalactic space. Researchers at the Los Alamos National Laboratory have come to the recognition that vast plasma streams extend through all cosmic space, combining all galaxies with networks of plasma streams.

Plasma streams that have galaxies at their node points

Milky Way look-alike
NGC 6744

ESA - Wide Field Imager view - CC BY 3.0

The plasma streams that have galaxies at their node points, with plasma being electric in nature, are inherently subject to electric resonance principles. For the Milky Way galaxy, which appears to be located at one of the node points between large galaxies, the resonance-characteristics in the plasma streams would become expressed in our galaxy as very long plasma-density cycles, with the length of the cycles corresponding to the length of the plasma streams.

The Andromeda galaxy

This means for us, that the closest galaxy to the Milky Way, named the Andromeda galaxy, which lays 2.5 million light years distant, appears to correspond to the 31 million-year resonance that we find modulated in the Milky Way, and the climate cycles on Earth.

The other connecting stream from the Milky Way

galaxy Messier 83

The other connecting stream from the Milky Way would be correspondingly longer. It would likely lead to the gigantic M83 galaxy that lays 15.2 million light years distant, and cause the 150-million-year cycle.

The resonance waves become overlaid

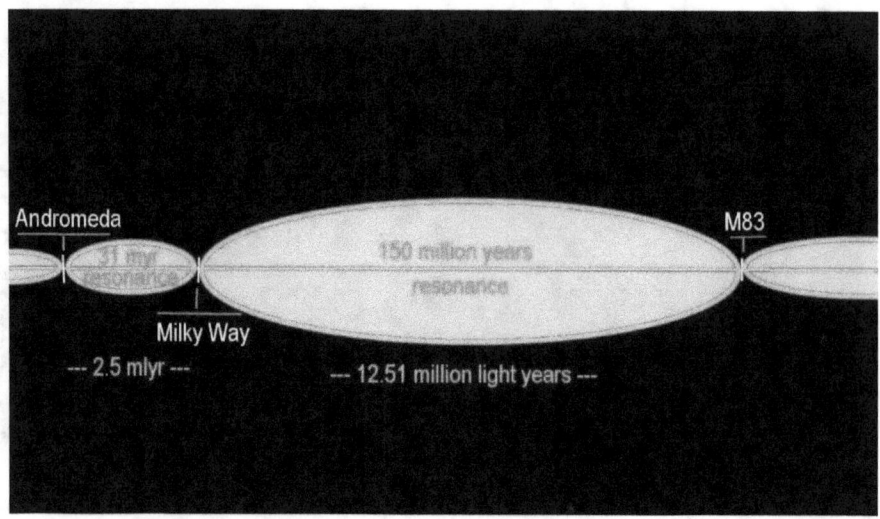

The resonance waves in both of these long-distance plasma streams coming together in the Milky Way, would give us the 31 and 150 million-years plasma density cycles respectively, that become overlaid and expressed as climate cycles on Earth.
The resulting evidence in historic climate cycles seems to tell us that the plasma source for our Sun, and for every sun in our galaxy, is located far-distant in intergalactic space, which the sink-effect of our sun, together with every star in our galaxy, draws from, which thereby keeps the entire plasma flow landscape in motion.

The fusion-sun sink process

However, the flow through the faucet depends on the pressure at the source. This means that the fusion-sun sink process depends for its functioning on the supply-line density, which is the plasma density in the galaxy.

The galactic plasma density is at a low point

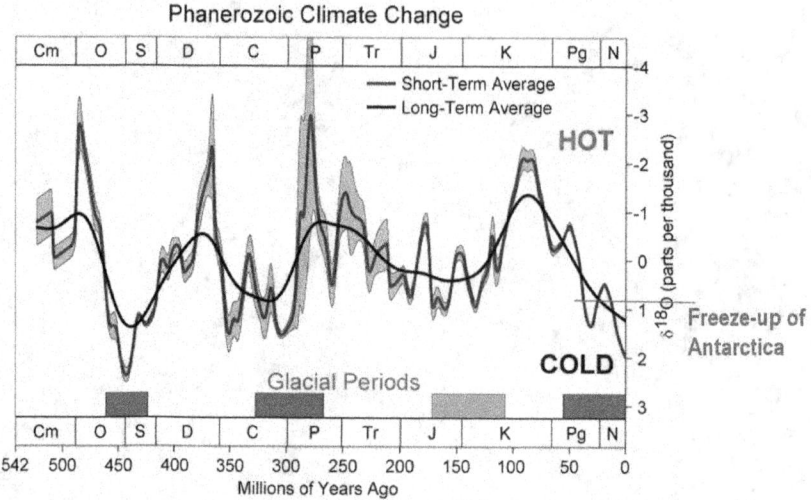

Phanerozoic Climate Change

Right now, the galactic plasma density is at a low point. The 31 and 150 million-years plasma density cycles are both at or near their lowest point. This combined low has brought us into the density zone were ice ages erupt, according to the dynamics of the local resonance cycles that are specific for our solar system. Here too, we see several cycles overlapping near their low point, with the result that our Sun will go inactive for the lack of plasma density.

When the sink effect draws more than the supply line holds

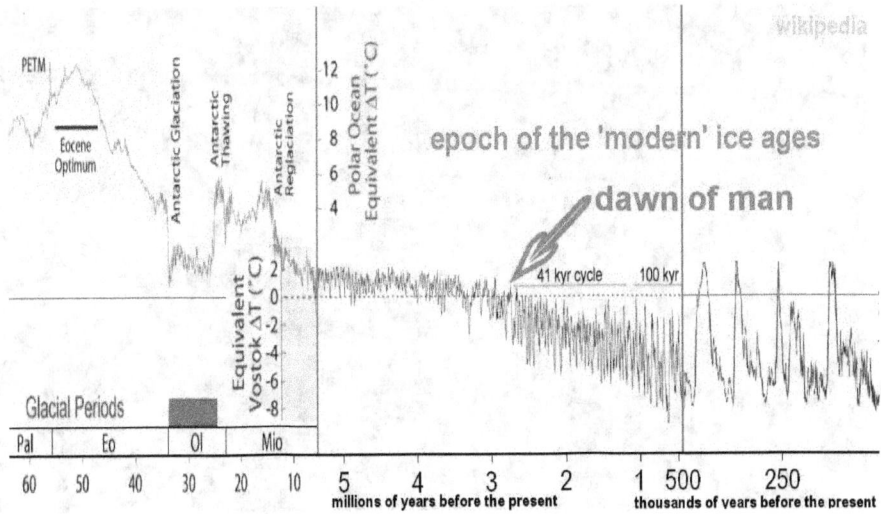

When the sink effect draws more than the supply line holds, the process becomes interrupted till the supply streams recover. Here the big Ice Age unfolds, with an inactive Sun, except for short periods, until the supply density becomes re-established. We are in this zone.

The universe is motivated by the plasma-sink process

The universe is motivated by the plasma-sink process. It affects everything.

The faucet as a sink

Since the need for the sink feature is not readily apparent, allow me to illustrate it once more, by comparing the plasma-flow system with a water supply system. Water supply systems typically have a faucet installed. When one opens the faucet, water flows. When this happens, the water in the system flows across the entire length of the supply line, all the way to the source that may be a lake high in the mountains. In such a system, the open faucet becomes a type of sink that enables the water to flow to where it is needed. With the faucet open, water flows throughout the entire length of the pipeline system. When the faucet is closed, nothing flows anywhere.

In the process of creating the uranium atom

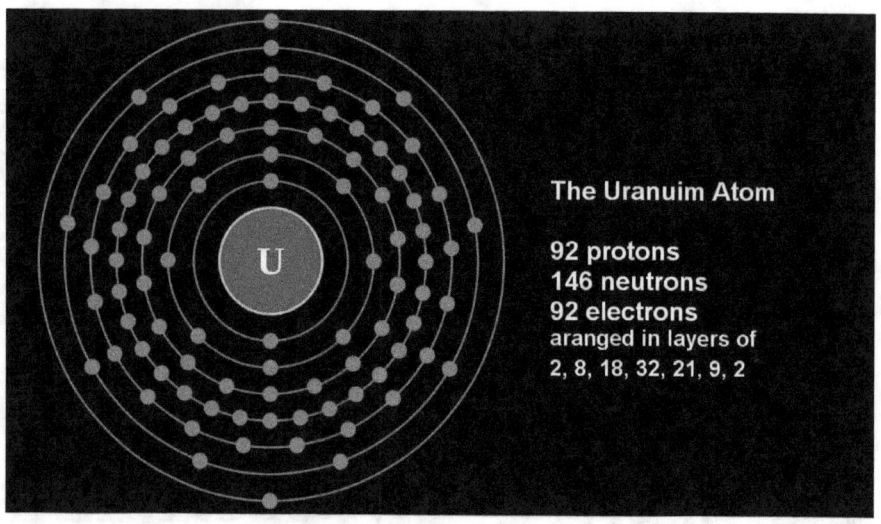

The Uranuim Atom

92 protons
146 neutrons
92 electrons
aranged in layers of
2, 8, 18, 32, 21, 9, 2

Now, let's look at an extreme example. In the process of creating the uranium atom, 238 protons were fused together to form the central nucleus for it. Before the protons were fused, they were a part of the plasma landscape and were repelling each other. After the atom is formed, the 238 protons suddenly are no longer a part of the electric landscape where they would add to the plasma pressure. The forming of the atom left behind a 'vacuum' as it were, in the plasma landscape. Of course, the vacuum is quickly filled with inflowing plasma. In this manner, a significant rate of flow is created in the movement of plasma, that becomes reflected throughout the entire system of plasma streams, reaching far back into interstellar space.

Interstellar plasma streams are set in motion by this consuming plasma fusion process that becomes a dynamic sink. This plasma sink is needed. Without a sink, no flow happens.

When plasma flows in interstellar space

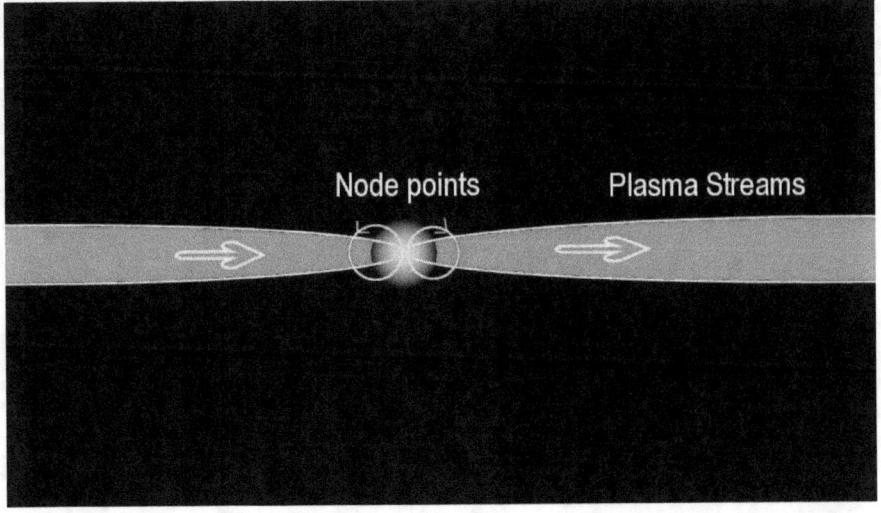

Now when plasma flows in interstellar space, the magnetic fields produced by the electric flow, draw the plasma streams closer together, by what is called the Lorentz force. Of course, when the plasma streams are drawn together, the electromagnetic pinching of them into an ever-smaller cross section increases the plasma density. The increased plasma density, in turn, increases the magnetic fields, which in turn pinches the current ever tighter. This self-feeding process forms the geometry of a bowl-type structure at the highest-density point. The magnetic fields become intensely tangled up at a point of high concentration, whereby the bowl-shaped end opens up and enables concentrated plasma to escape. The node points would logically occur near a plasma sink, which motivates the plasma flow in the first place.

Assume that the plasma sink is our Sun

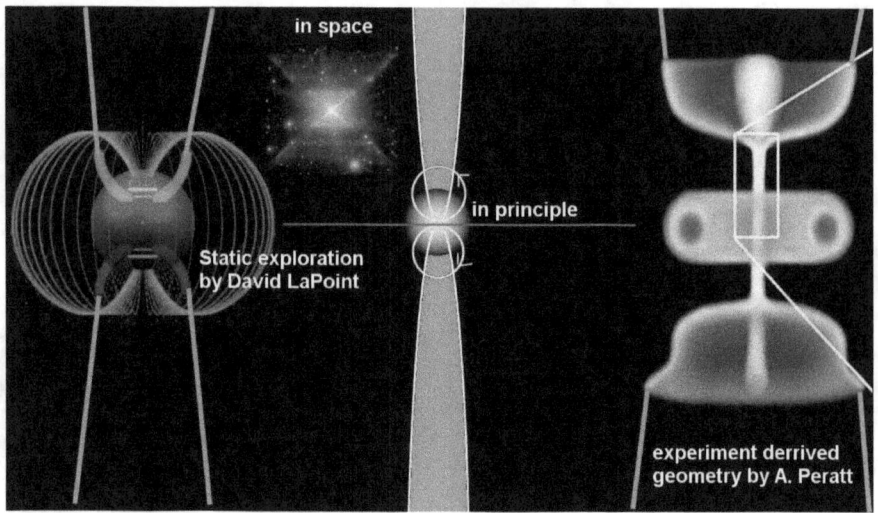

Let's assume that the plasma sink is our Sun, as the center of the system. In this case the bowl shaped magnetic fields would be focused onto our Sun.

In the Red Square Nebula

Normally, interstellar plasma streams are not visible, because the plasma particles are too small to be visible, but in a few extremely high-powered examples, such as in the Red Square Nebula, some parts of the bowl-shaped geometry do become visible in the areas of the highest plasma concentration and atomic concentration, where the dance of the electrons in plasma interacts with the synthesized atoms.

Contrary to general perception, nebulas are not the remains of exploded stars, but are high-power electromagnetic structures. In the case of Red Square Nebula, a number of unique features of the high-power electromagnetic bowl-type structure become visible.

The plasma researcher David LaPoint

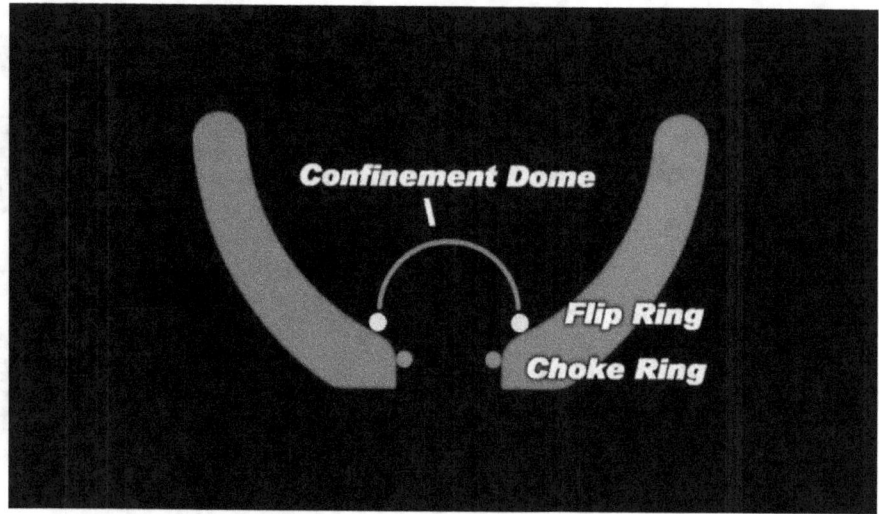

The plasma researcher David LaPoint has extensively explored the nature of the magnetic fields of bowl-shaped magnetic structures, which he has termed The Primer Fields. He identified three unique magnetic features. He has identified what he calls a confinement dome, a flip ring, and a choke ring.

The flip ring is a magnetic ring

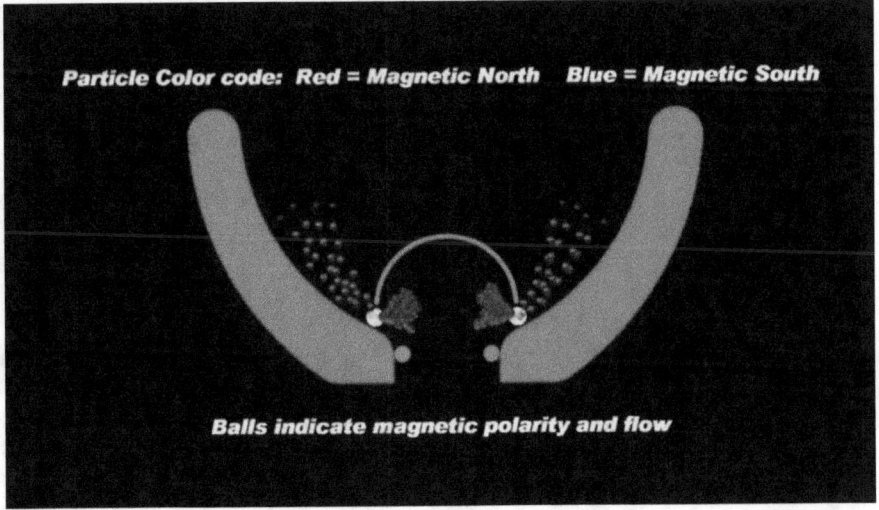

Particle Color code: Red = Magnetic North   Blue = Magnetic South

Balls indicate magnetic polarity and flow

The flip ring is a magnetic ring that flips the incoming plasma stream upwards under the confinement dome where plasma becomes intensely concentrated. The concentrated plasma, becomes then focused downward by the choke ring that focuses the concentrated plasma onto a sun that acts as a sink for the plasma flow. All three features are visible in the Red Square Nebula.

The features that we see in the Red Square Nebula

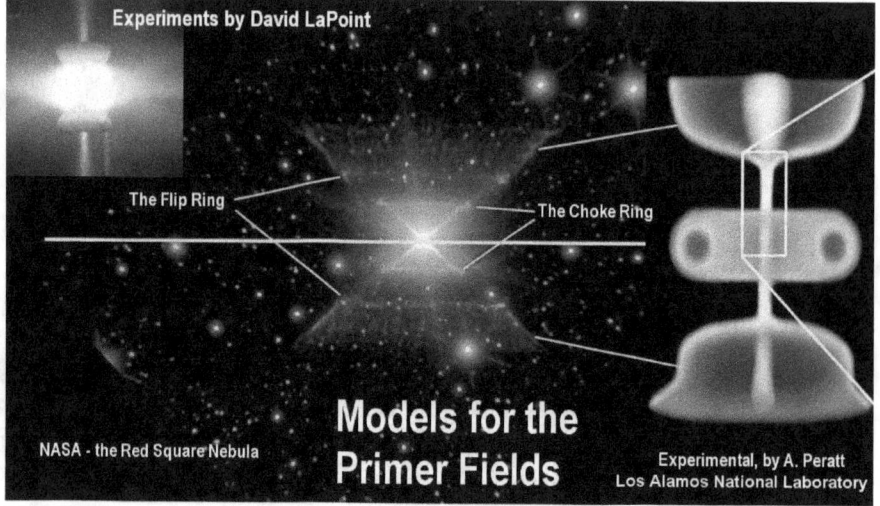

The features that we see in the Red Square Nebula correspond with what David LaPoint has been able to replicate in the laboratory. The features also match in principle the resulting plasma shapes observed in high-energy plasma flow experiments conducted at the Los Alamos National Laboratory, by Antony Paratt, director of experiments.

The flow of electricity is critical

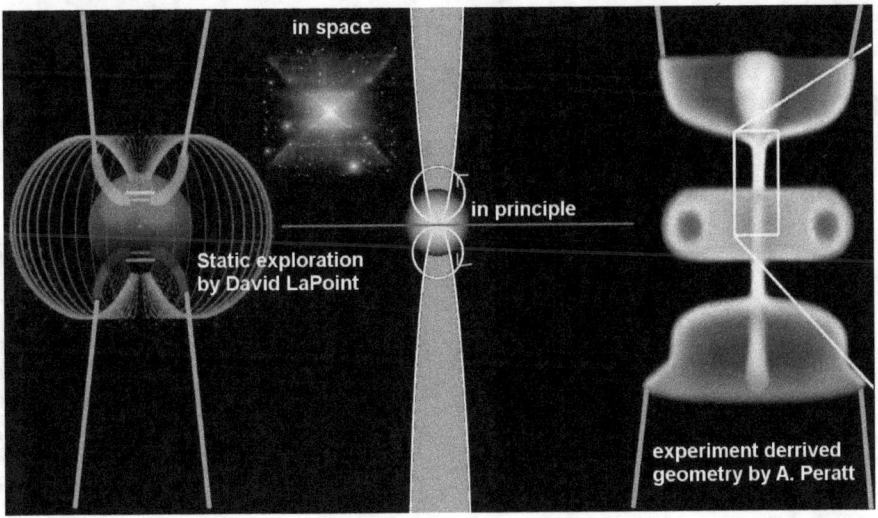

In electrodynamic systems, the flow of electricity is critical. It is exclusively the flow of electricity that creates magnetic fields. No other cause for magnetic effects exists.

By the resulting magnetic fields

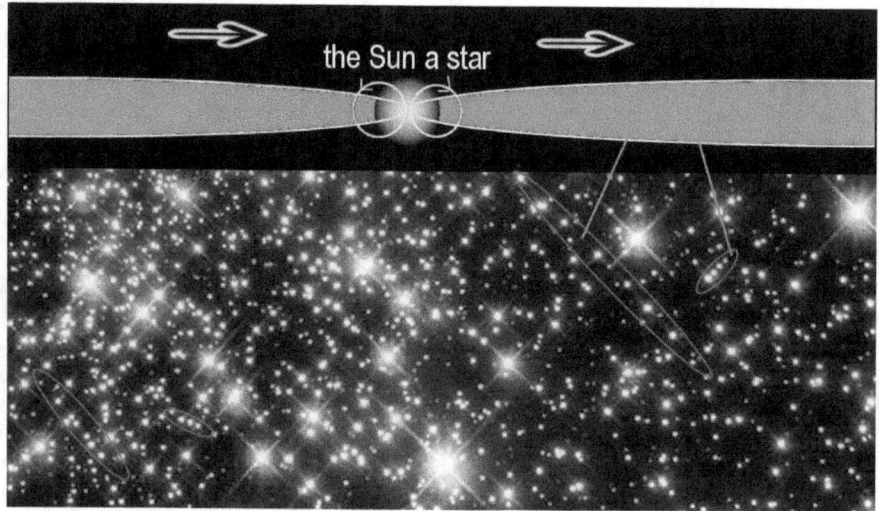

By the resulting magnetic fields, the electric plasma streams become aligned, they become pinched together, and become increasingly concentrated by the magnetic fields in such a manner that a high density plasma sphere forms around a sink object, like our Sun is.

Plasma surrounding our Sun becomes so dense

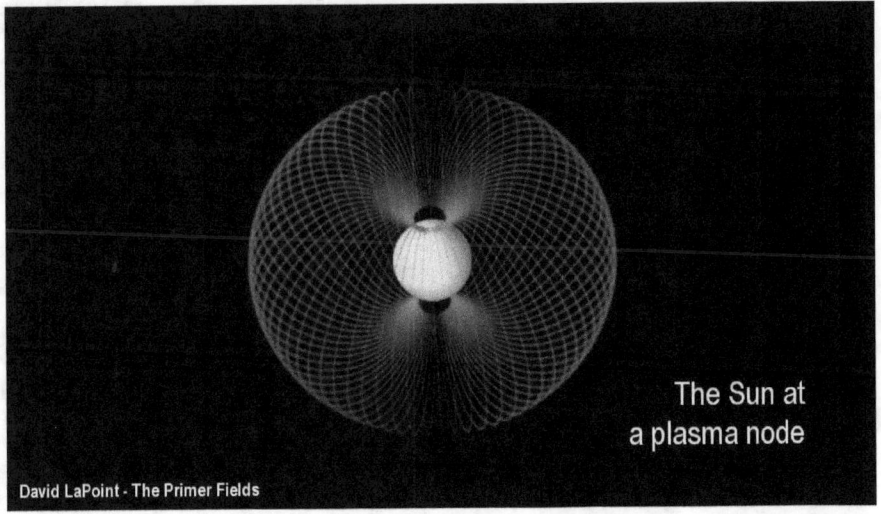

The Sun at
a plasma node

David LaPoint - The Primer Fields

By this process the plasma surrounding our Sun becomes so dense
that electric fusion reactions can, and do, occur at the Sun's surface
in intensely concentrated plasma, magnetically confined by the
primer fields.

This is what causes our Sun to become a sink

Theoretical, by David LaPoint

Experimental,
by David LaPoint

This is precisely what causes our Sun to become a sink for interstellar plasma, which it must be, in order to motivate the flow of plasma into it?

The Sun is seen as a vast sea of granular cells

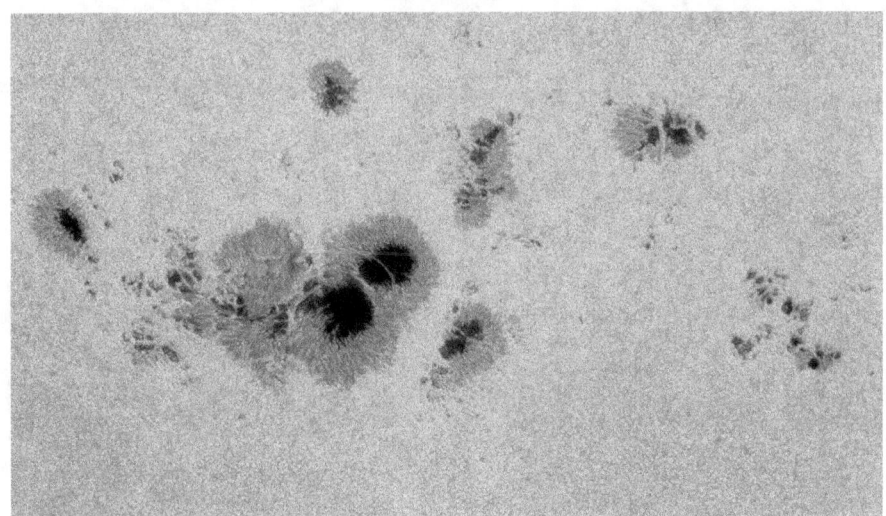

For a closer look, we need to look at the Sun's surface.
When the Sun is observed with an optical telescope, the Sun is seen as a vast sea of granular cells, which sometimes break down, leaving sunspots in the wake.

The graduals are cells of Primer Fields

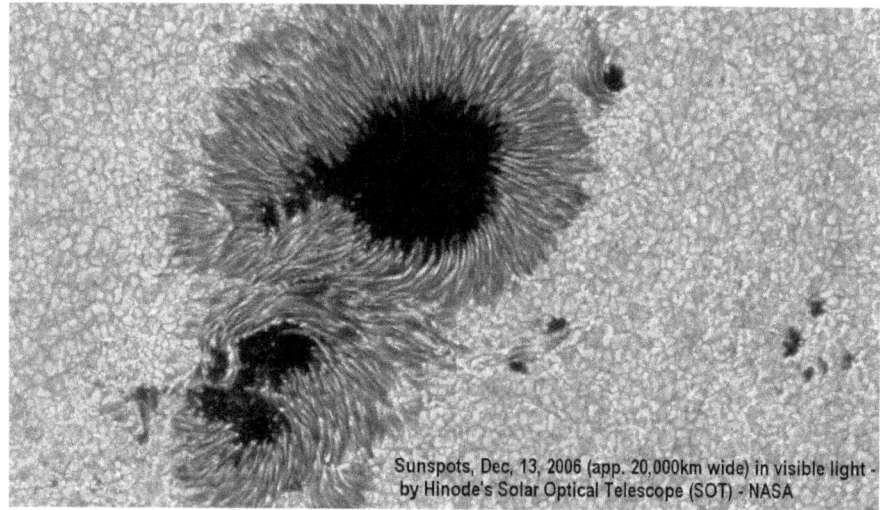

Sunspots, Dec, 13, 2006 (app. 20,000km wide) in visible light - by Hinode's Solar Optical Telescope (SOT) - NASA

Since the Sun is powered on its surface by electric plasma interaction, the graduals are themselves cells of Primer Fields in operation, on a relatively 'small' scale. The cells are typically a thousand kilometers across.

The Sun is a vast sea of cellular primer field structures

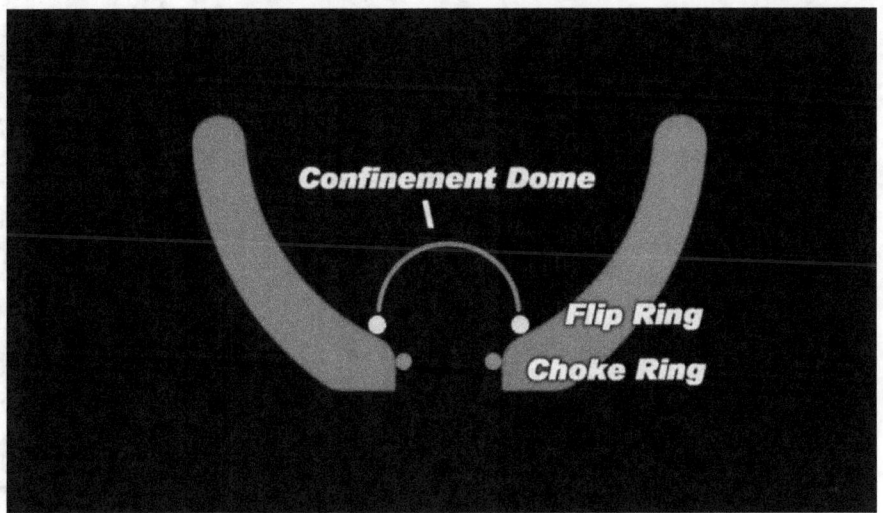

This means that the surface of the Sun is a vast sea of cellular primer field structures, all operating side by side, with each having its own flip ring and confinement dome.

Here again, the inflowing plasma

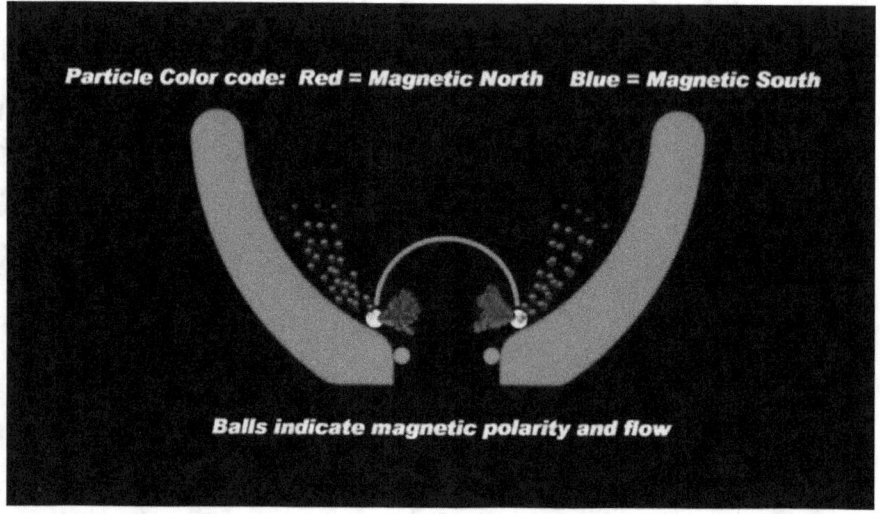

And here again, the inflowing plasma, that is at this stage highly concentrated around the Sun, becomes even more extremely concentrated under each cell's confinement dome.

Nuclear fusion causes the Sun to act as a plasma sink

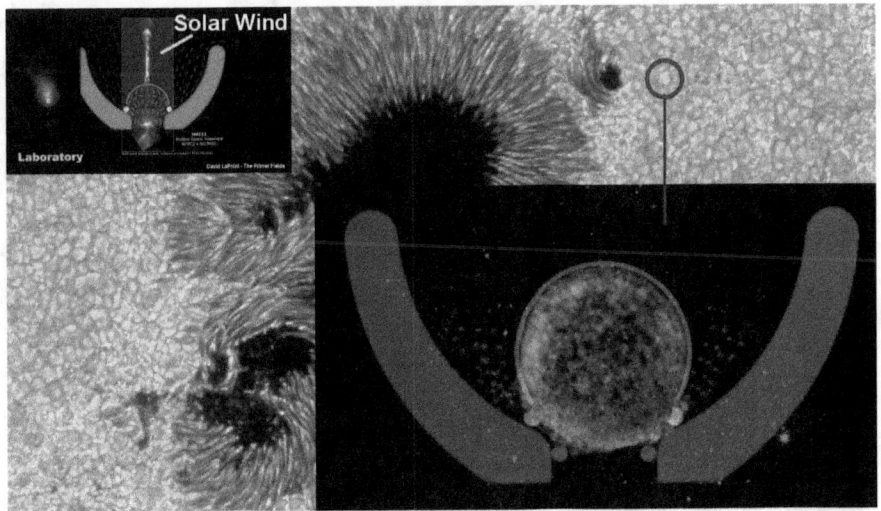

In this case the plasma becomes so immensely concentrated, and accelerated, under the confinement dome, that nuclear fusion occurs. As I said before, the resulting nuclear fusion process causes the Sun to act as a plasma sink.

Without the fusion process on the surface of the Sun, which synthesizes all natural atoms, the Sun could not be powered with external energy. The reactions simply would not occur. Nothing would flow. However, when atoms are formed, the previously electrically active plasma particles suddenly become so tightly packed and perfectly balanced that they become electrically neutral. As a consequence, they cease to exist in the electric landscape.

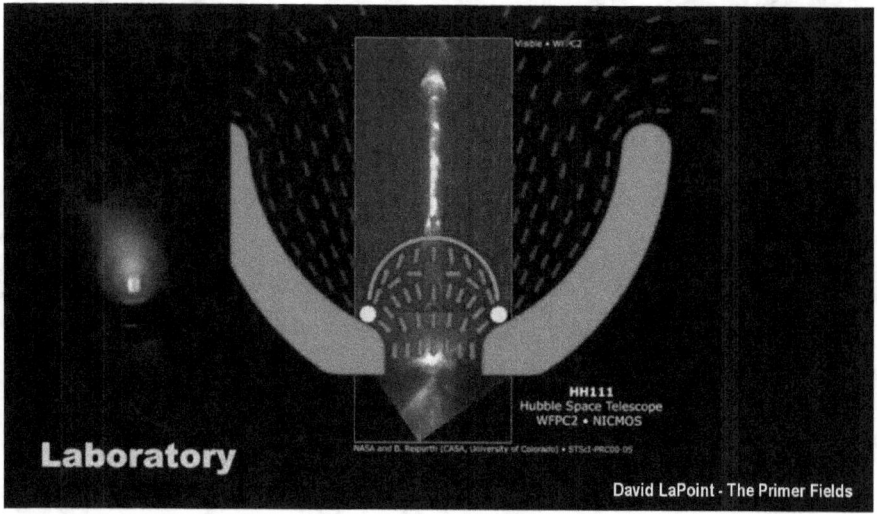

By being no longer tied to anything, the synthesized atoms flow away with the solar wind. By this process of electric atomic synthesis, a powerful sink effect is created, in which, in addition, all the natural elements that exist, that the planets are made of, were and are created by the process.

The synthesis extends far beyond the helium fusion stage

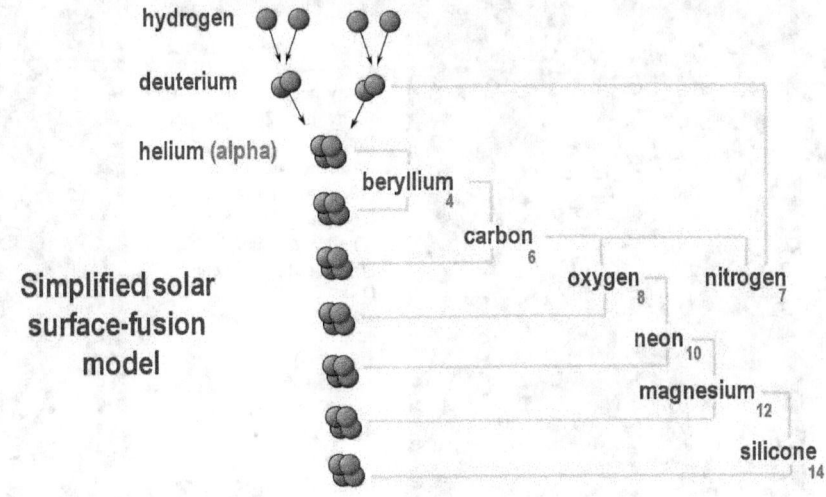

hydrogen

deuterium

helium (alpha)

beryllium 4

carbon 6

oxygen 8   nitrogen 7

neon 10

magnesium 12

silicone 14

Simplified solar
surface-fusion
model

Instead of the p-p chain of fusion that is theorized for the internally heated Sun, which ends at the helium-4 stage, no such limit is inherent in the plasma powered surface fusion process. The synthesis extends far beyond the helium fusion stage.

In the remarkably close agreement

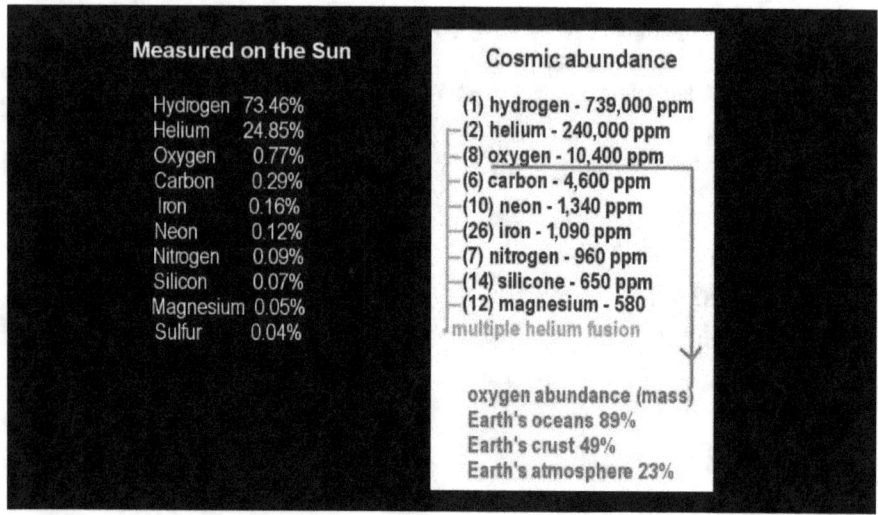

The evidence that a wide range of surface fusion is happening, is found in the remarkably close agreement of the measured element ratios on the surface of the Sun, shown on black, with the known cosmic abundance ratio for the same elements in the solar system, shown on white.

The heavier elements, past the helium stage

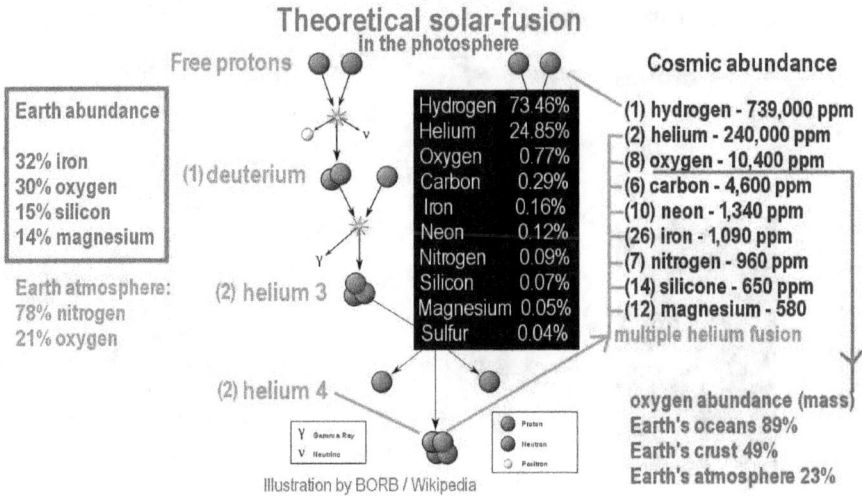

Theoretical solar-fusion
in the photosphere

Free protons

Earth abundance
32% iron
30% oxygen
15% silicon
14% magnesium

Earth atmosphere:
78% nitrogen
21% oxygen

(1) deuterium

(2) helium 3

(2) helium 4

Hydrogen	73.46%
Helium	24.85%
Oxygen	0.77%
Carbon	0.29%
Iron	0.16%
Neon	0.12%
Nitrogen	0.09%
Silicon	0.07%
Magnesium	0.05%
Sulfur	0.04%

Cosmic abundance

(1) hydrogen - 739,000 ppm
(2) helium - 240,000 ppm
(8) oxygen - 10,400 ppm
(6) carbon - 4,600 ppm
(10) neon - 1,340 ppm
(26) iron - 1,090 ppm
(7) nitrogen - 960 ppm
(14) silicone - 650 ppm
(12) magnesium - 580
multiple helium fusion

oxygen abundance (mass)
Earth's oceans 89%
Earth's crust 49%
Earth's atmosphere 23%

Y Gamma Ray
V Neutrino

Proton
Neutron
Positron

Illustration by BORB / Wikipedia

The heavier elements, past the helium stage, should therefore not be present in the atmosphere of the Sun at all. None should be there. Still, they have all been measured there. Their presence, that shouldn't be, invalidates the internal-fusion theory still further.

95

The presence of 'heavy' elements in the solar atmosphere

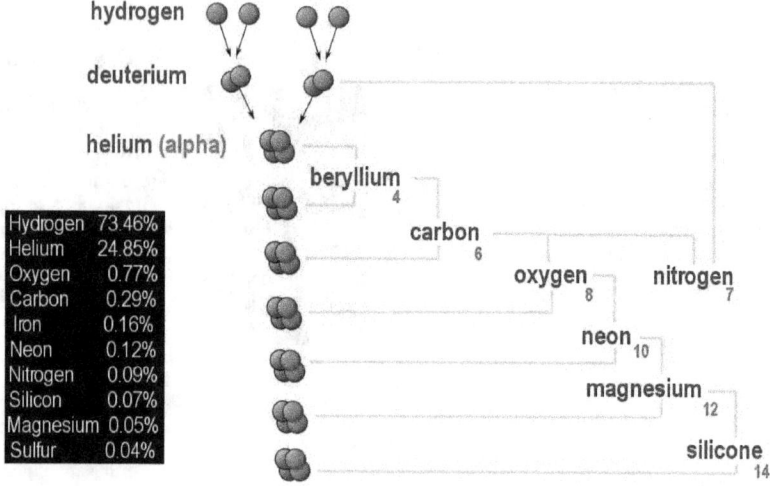

Hydrogen	73.46%
Helium	24.85%
Oxygen	0.77%
Carbon	0.29%
Iron	0.16%
Neon	0.12%
Nitrogen	0.09%
Silicon	0.07%
Magnesium	0.05%
Sulfur	0.04%

The presence of 'heavy' elements in the solar atmosphere proves the electric surface-fusion concept, which alone enables atomic synthesis to be possible on the Sun, in a process in which atoms are fused directly from plasma. This includes all the known natural elements in the periodic table.

If all of these heavy elements were synthesized in the core of the Sun, these elements would have accumulated there and would have become the core, a very heavy core, which the Sun does not have, and cannot have.

This type paradox is not found on the external fusion platform.

Evident by the existence of noctilucent clouds

Noctilucent clouds over Kuresoo bog, Viljandimaa, Estonia, app. 75-85 Km high - wikipedia

That strong fusion synthesis is happening on the surface of the Sun, is visibly evident on Earth. It is evident by the existence of noctilucent clouds high in the stratosphere. We see rivers of water vapor present there, frozen into ice crystals, that were evidently not lifted up from the surface 80 kilometers high into the stratosphere, but came instead directly from the Sun, carried by the solar winds.

Cold fusion drives the universe

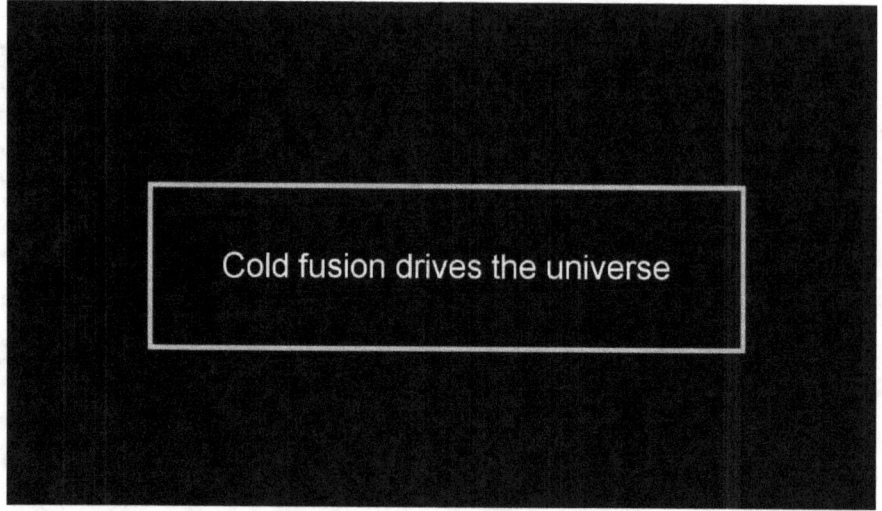

Cold fusion drives the universe

Electric fusion is cold fusion

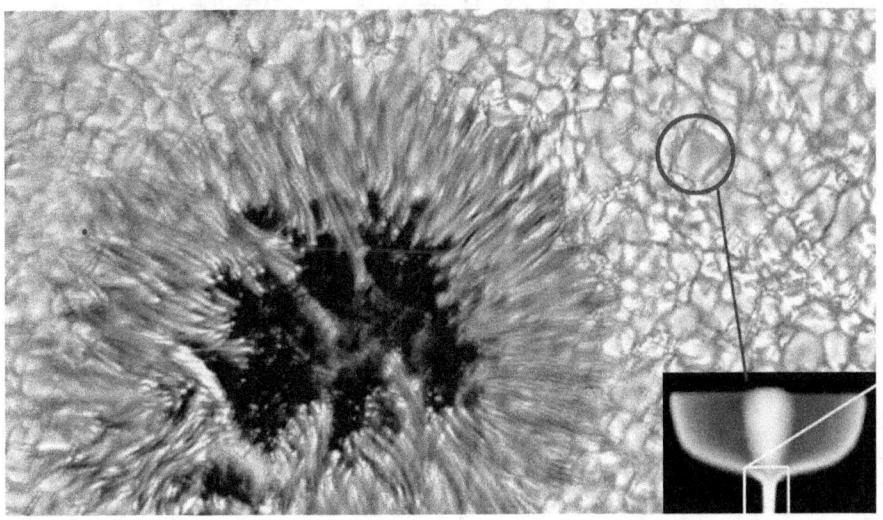

Electric fusion on the surface of the Sun is cold fusion. It is enabled by magnetic plasma compression and is maintained by the sink effect of the fusion itself. The heating of the photosphere is a subsequent phenomenon. It is comparatively minuscule phenomenon, at 5,505 degrees Celsius.

All the nuclear fusion experiments on Earth are based on the illusion that the fusion process inside the Sun, which the experiments aim to replicate, require those 15 million degrees that are theorized to exist in the Sun for the internal heating that is deemed to so violently agitate the atoms that they become forced to fuse. For this reason, all fusion experiments on Earth, are hot fusion experiments, even though hot fusion doesn't actually happen on the Sun, and likely nowhere in the universe either.

Cold nuclear fusion is the fusion that has created all the elements that the planets are made of. The atoms in the Universe are not the result of the mythical, primordial Big Bang, but are the product of cold-powered electric fusion that is motivated exclusively by the electric force and is maintained by the fusion-process itself, as it creates an electric sink where plasma electricity is being bottled up and neutralized into packages of electrically neutral atoms.

Atomic elements are extremely rare in the universe

Group → Period ↓	1	2	3	4	5	6	7	8	9	10	11	12	13	14	15	16	17	18
1	1 H																	2 He
2	3 Li	4 Be											5 B	6 C	7 N	8 O	9 F	10 Ne
3	11 Na	12 Mg											13 Al	14 Si	15 P	16 S	17 Cl	18 Ar
4	19 K	20 Ca	21 Sc	22 Ti	23 V	24 Cr	25 Mn	26 Fe	27 Co	28 Ni	29 Cu	30 Zn	31 Ga	32 Ge	33 As	34 Se	35 Br	36 Kr
5	37 Rb	38 Sr	39 Y	40 Zr	41 Nb	42 Mo	43 Tc	44 Ru	45 Rh	46 Pd	47 Ag	48 Cd	49 In	50 Sn	51 Sb	52 Te	53 I	54 Xe
6	55 Cs	56 Ba		72 Hf	73 Ta	74 W	75 Re	76 Os	77 Ir	78 Pt	79 Au	80 Hg	81 Tl	82 Pb	83 Bi	84 Po	85 At	86 Rn
7	87 Fr	88 Ra		104 Rf	105 Db	106 Sg	107 Bh	108 Hs	109 Mt	110 Ds	111 Rg	112 Cn	113 Uut	114 Fl	115 Uup	116 Lv	117 Uus	118 Uuo

Lanthanides	57 La	58 Ce	59 Pr	60 Nd	61 Pm	62 Sm	63 Eu	64 Gd	65 Tb	66 Dy	67 Ho	68 Er	69 Tm	70 Yb	71 Lu
Actinides	89 Ac	90 Th	91 Pa	92 U	93 Np	94 Pu	95 Am	96 Cm	97 Bk	98 Cf	99 Es	100 Fm	101 Md	102 No	103 Lr

It is not unreasonable therefore, to assume, that all the atomic elements in the solar system, which the planets are made of, which together make up fourteen-one-hundredth of a percent of the visible mass of the solar system, where all synthesized on the surface of the Sun in a nuclear fusion process that creates atomic structures. The remaining 99.86% of the mass of the solar system is plasma located in the Sun, and this ratio does not even include the mass of the plasma in the Primer Fields and in the solar winds, and so on.

The extremely high plasma-mass ratio tells us that atomic elements are extremely rare in the universe, even though they are the heart of the world in which we live.

Every element of our world was synthesized on the Sun over time. Most of this synthesis occurred during the epoch in which the Sun itself was forged in the high-density plasma-landscape near the center of the galaxy. This means that the synthesizing system that produces the vast abundance of different atomic elements that make up our world, must necessarily be enormously efficient. The great efficiency is rooted in the dynamics of electrically motivated cold fusion that creates its own sink.

The cold-fusion process in the Sun is simple

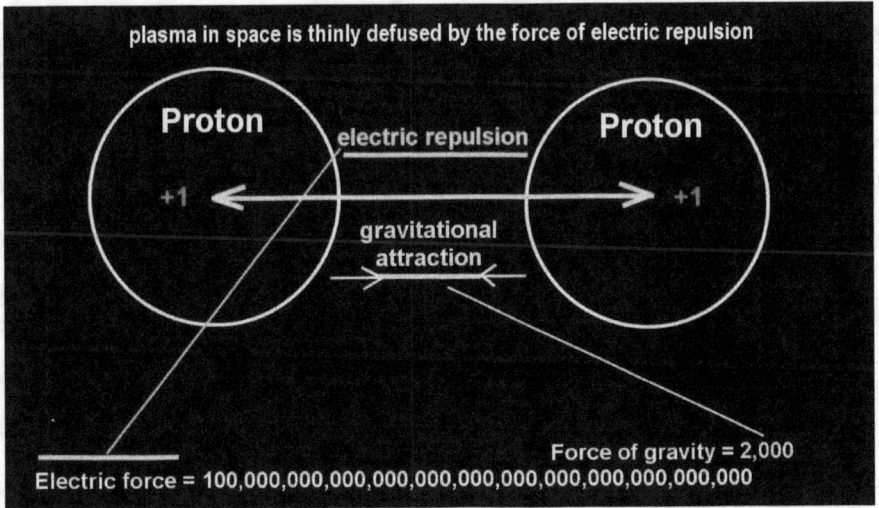

The cold-fusion process in the Sun is simple. All protons repel each other strongly by the electric force that is one of the basic forces of the universe. Each proton also packs a substantial mass that is roughly 2,000 times greater than the mass of an electron.

When energy is invested

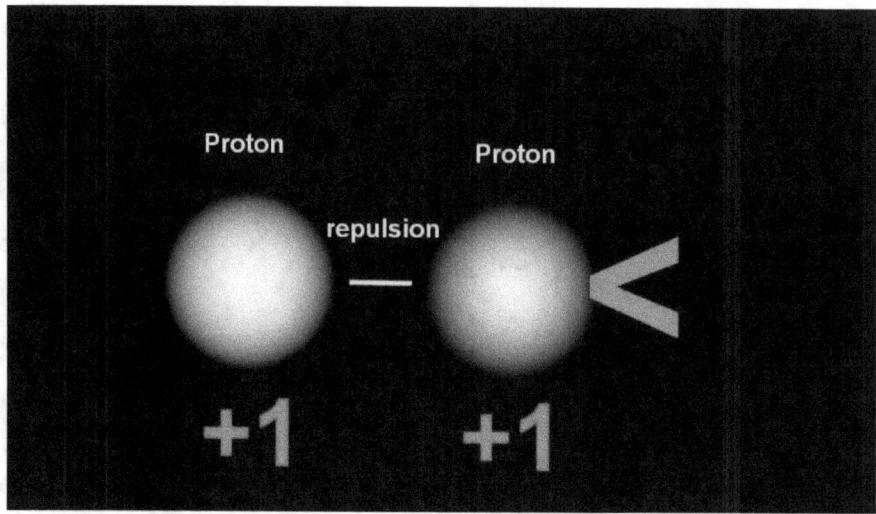

When energy is invested into putting the mass of a proton into motion, and this kinetic energy becomes great enough to overcome the electric repulsion, the protons will simply snap together. They will fuse.

As the protons fuse

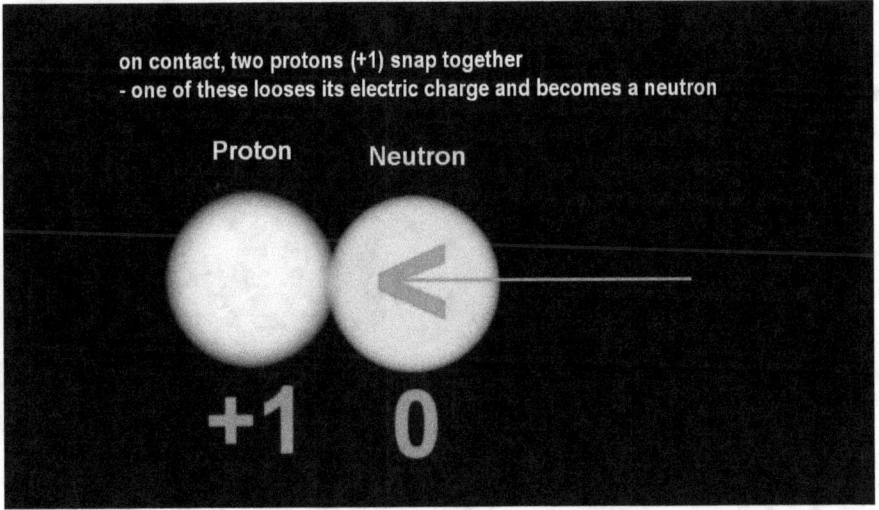

As the protons fuse, some of the kinetic energy becomes 'loaded' into one of the fusing protons. The proton becomes transformed thereby. It becomes transformed by the process, into a neutron. The resulting neutron carries no electric charge. As the result of this fusion process, the electric field of one of the joining protons simply ceases to exist.

The resulting field reduction

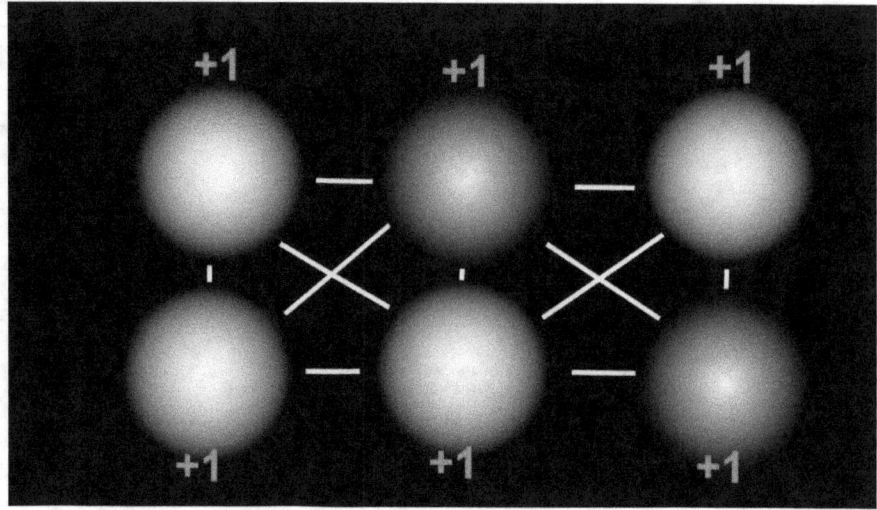

The resulting field reduction has an immensely dramatic effect on the plasma stream where the protons repel one another, as in the case shown here.

The dynamics create an imploding effect

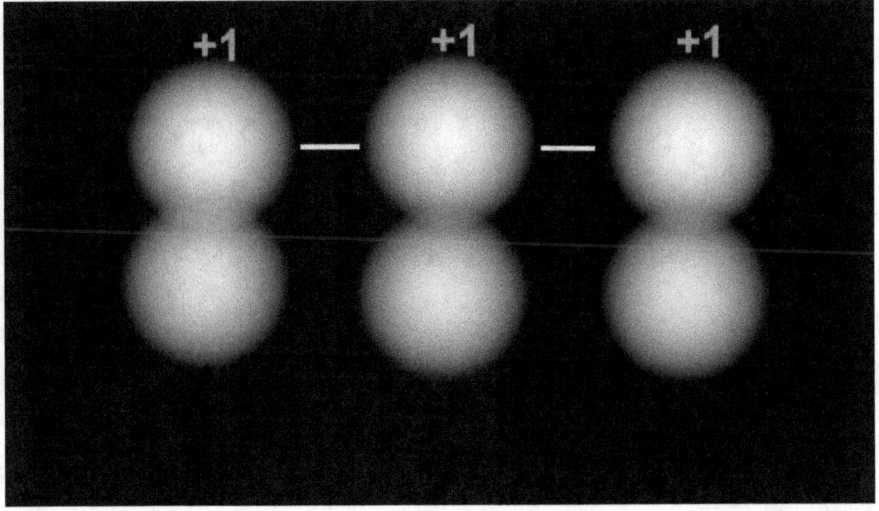

In this manner, the overall repelling effect that affects the plasma density, is much more-dramatically reduced. The dynamics create an imploding effect.

The resulting 'shock' of the vanishing electric fields and the lost repulsion effect, creates a partial 'void' in the plasma pressure, which the adjoining protons are propelled to fill. Thus, the fusion effect creates, by the electric force, some highly energized movement in the plasma stream towards the Sun. The plasma movement, of course, further aids the plasma compression, and thereby aids the fusion process.

The resulting voids accelerates the remaining plasma

The resulting voids accelerates the remaining plasma, just as if one had opened a water-faucet more fully. This process is self-escalating.

The more protons are joined, the more neutrons are created, and the greater the void in the repelling electric field becomes, which in turn accelerates the process in leaps and bounds. This means that the faucet is effectively opened evermore, the stronger the process flows. Nor does it end here.

As more protons are being used up

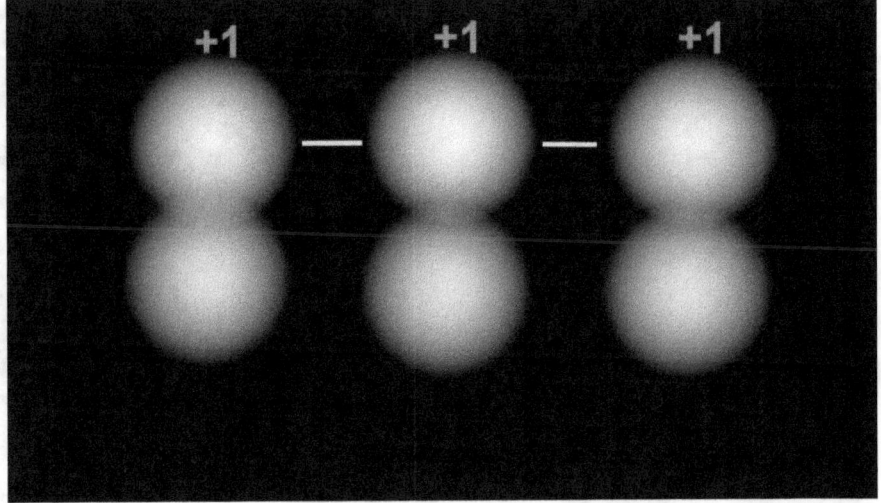

As more protons are being used up, by becoming neutrons that have no electric effect, the ratio between the remaining electrons and protons the in the plasma stream, suddenly doubles, and in some cases triples.

As the fused proton and neutron clusters are joined

As the fused proton and neutron clusters are joined to one-another themselves, forming ever larger clusters, the overall electron density surrounding the larger clusters is thereby increased evermore, to the point that the electrons, too, become highly energetic, and become bound up by the now increasing electric energy strength of the accumulating nuclei. When this happens, the result is even more dramatic.

The entire electric scene becomes a great void

When the electric charge of the electrons balances the charge of the remaining protons, the entire electric scene becomes a great void. All protons and electrons in the entire atomic construction loose their overall electric field potential as if they had vanished from the universe.

The void is so great that substantial plasma streams are drawn into it, like into a bottomless pit. In this manner, the electric nuclear-fusion synthesis creates in effect a huge bottomless pit that eats up electric fields in plasma. This bottomless pit sucks up plasma, by which atomic elements emerge.

In comparison, the waterline carries the plasma streams

The faucet
as a sink

wikipedia

In comparison, the waterline carries the plasma streams. The faucet
is the fusion process that the plasma flows into. Out of the faucet
comes a steam of assembled atoms that are electrically neutral and
dissipate into the landscape. In the process the faucet gets
exceedingly hot. The energy that drives the system comes from the
active fusion process, enabled by the plasma-sink principle.

The same principle applies

Grande Dixence Dam
935 feet tall - Switzerland
wikipedia

The same principle applies here that we utilize on Earth to drive hydroelectric generating plants, powered by the pressure differential between the reservoir level, and the discharge level.

Energy is carried away by electric transmission lines

From hydro-electric plants, the generated energy is carried away by electric transmission lines.

The process heat is radiated as light

From the Sun, the process heat is radiated as light and heat, enabling life to be.

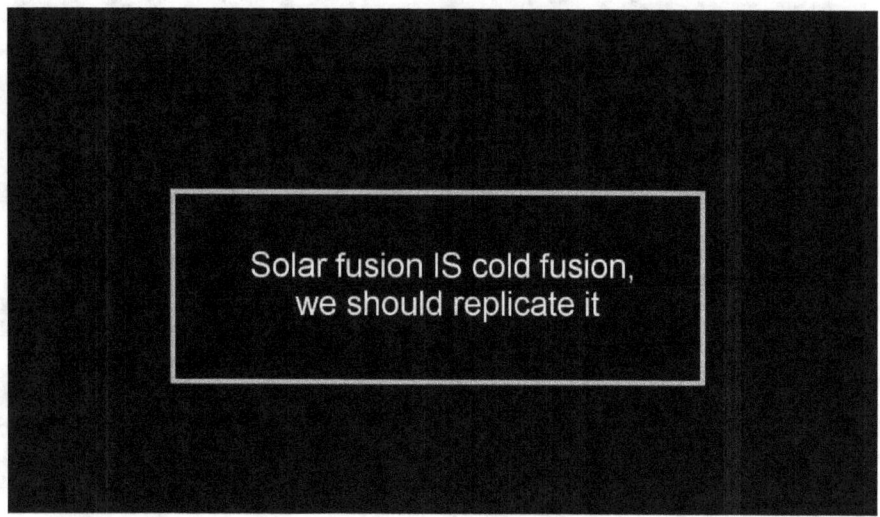

Solar fusion IS cold fusion, we should replicate it
Cold fusion, as we see the principle applied on the Sun, is the most efficient energy producer in the universe. It is anti-entropic in nature, with a resource that is self-renewing.

Cold fusion is efficient as a solar process

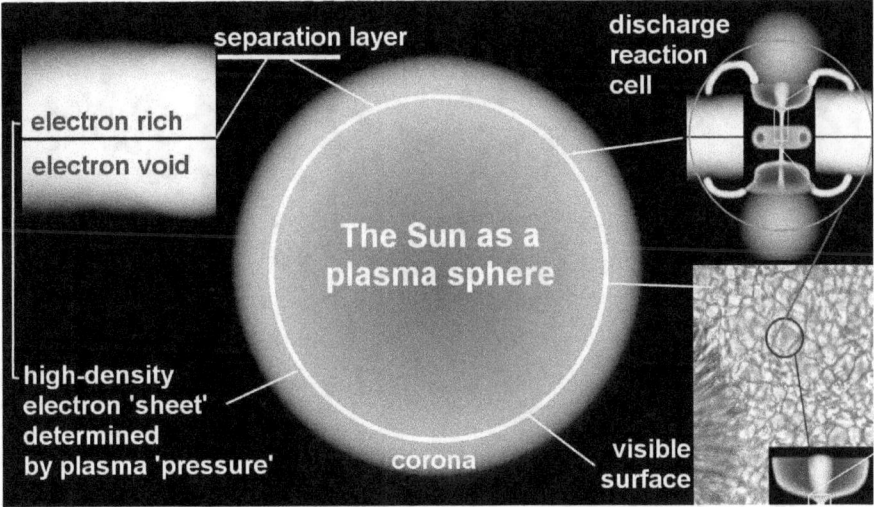

Cold fusion is efficient as a solar process, because its active component is self-accelerating. Thereby, large amounts of energy and atomic elements are generated with relative 'ease,' on the solar surface. Nothing is forced there. Everything happens naturally. The 'sink' effect is so efficient in accelerating plasma that the millions of degrees in thermal energy that the fusions experiments on Earth require, to force atoms to fuse together, are not required on the Sun. The Sun operates with a 'cold' fusion process in which thermal and light energy is produced as a secondary effect by the mechanics of the process.

Artificial nuclear-fusion power processes fail

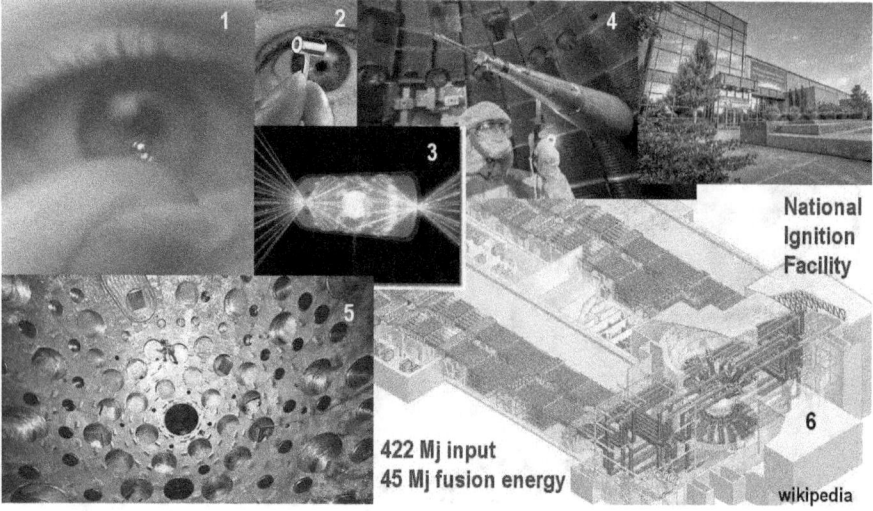

National Ignition Facility

422 Mj input
45 Mj fusion energy

wikipedia

All of the current, artificial nuclear-fusion power processes fail, because they fail to utilize the critical 'sink' principle that motivates cold fusion. They employ enormously large external forces in an effort to break the electric repulsion barrier that develops between approaching atomic nuclei, once the electron shells are penetrated.

The kind of agitation of atoms that breaks this barrier, requires enormous energies as input to make the thing possible. And even with all that, efficient fusion has not yet been achieved, not even with such giant efforts as are made at the National Ignition Facility, which aims for inertial fusion by means of light-energy compression. The effort has failed so far, in spite of the most ideal fusion fuel being used for the experiment.

The easiest-to-fuse fusion fuel

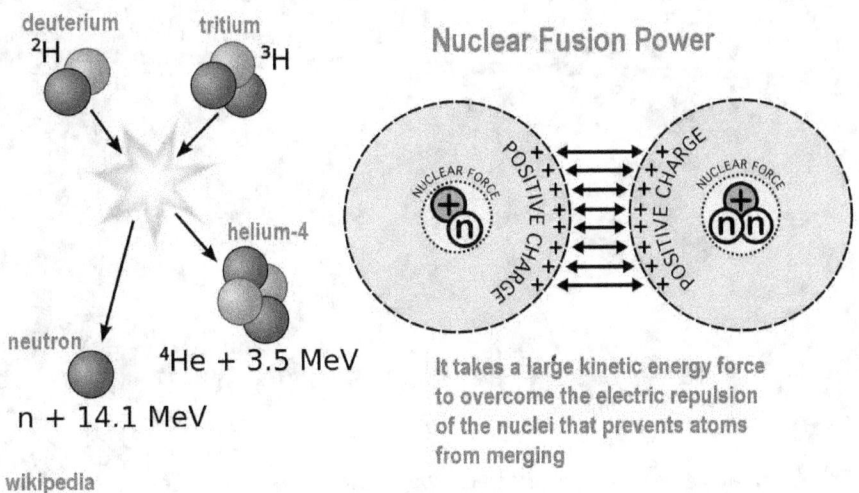

deuterium
^{2}H

tritium
^{3}H

Nuclear Fusion Power

helium-4

neutron

^{4}He + 3.5 MeV

n + 14.1 MeV

wikipedia

It takes a large kinetic energy force to overcome the electric repulsion of the nuclei that prevents atoms from merging

The easiest-to-fuse fusion fuel is made up of two types of over-built atomic isotopes, deuterium and tritium, that are violently forced to connect with each other. When they do connect and fuse, one of the overbuilt isotopes releases a redundant neutron that is not required for the end product. When this happens, its previously invested binding energy becomes liberated. This liberated energy is intended to be used to power electricity generating systems. This, in short, is the dream scenario.

Attempted power production processes all invariably fail

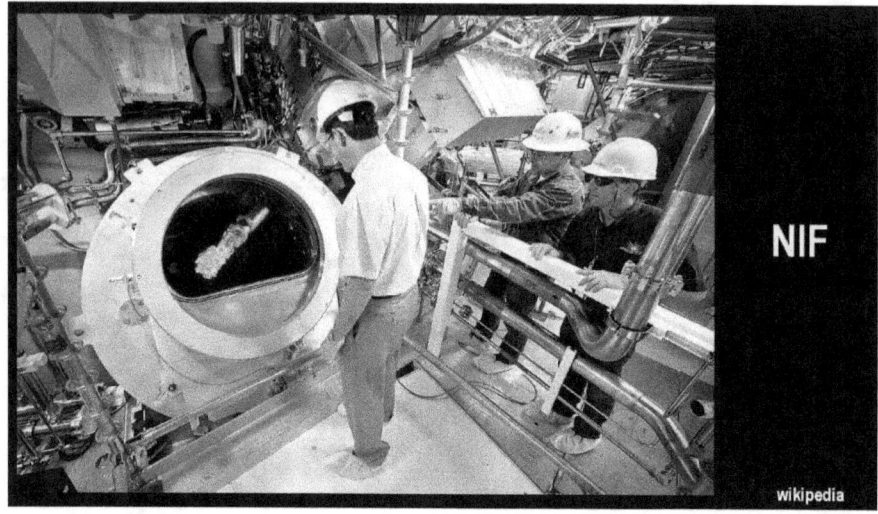

The attempted power production processes all invariably fail, because the intended fusion process does not aim for the natural 'sink' effect, but forces results with the injection of brute agitation for which enormous energies are expended to generate the agitation. Consequently, it takes radically more energy as input, to operate such a fusion process, than the result gives back.
The inherent inefficiency that results from substituting the natural 'sink' principle, with brute force, renders nuclear-fusion power production on Earth, a hopeless proposition. This basic failure in principle stands in addition to all the other problems associated with the projects that render the hoped-for success an empty dream.

Natural solar processes are not yet utilized for power production

Magnetic Confinement Fusion - Tokamak

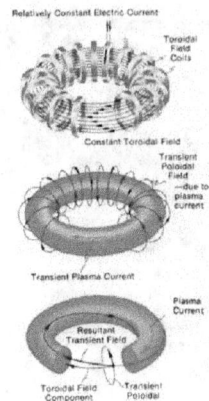

When it is being said that terrestrial nuclear-power projects aim to replicate the process that powers the Sun, a fundamental error is incurred, because what is being replicated in experiments, does not occur, naturally, either on the Sun, nor anywhere else, as a process. The actually solar processes are infinitely more efficient than what the experiments aim for. The natural solar processes that are in operation on the Sun, are not yet utilized for power production anywhere on Earth, experimentally or otherwise, for the simple reason that they are deemed not to exist.

This does not mean that the real solar-energy processes

This does not mean that the real solar-energy processes will never be replicated on Earth? Not necessarily.

Plenty of evidence exists

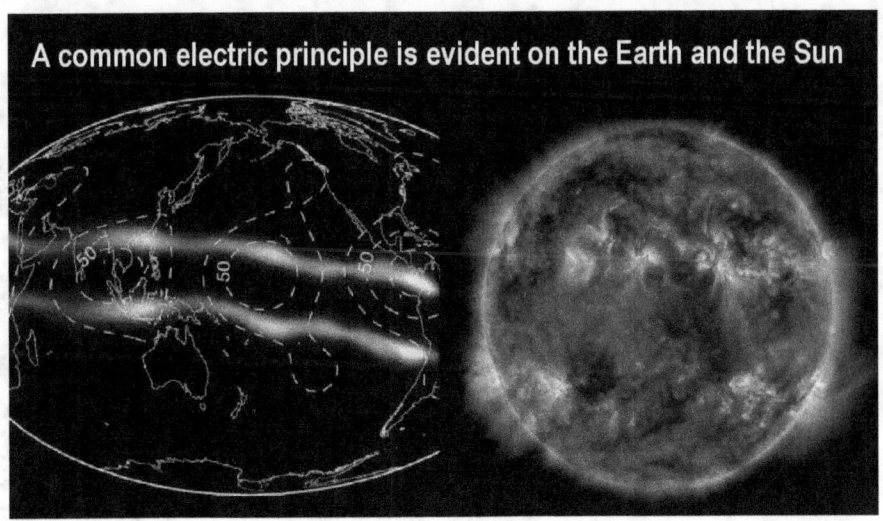

A common electric principle is evident on the Earth and the Sun

Plenty of evidence exists that cosmic electric plasma surrounds the Earth, in a pattern that is also visible on the Sun.

A plasma-flow pattern is 'visible' in the ionosphere

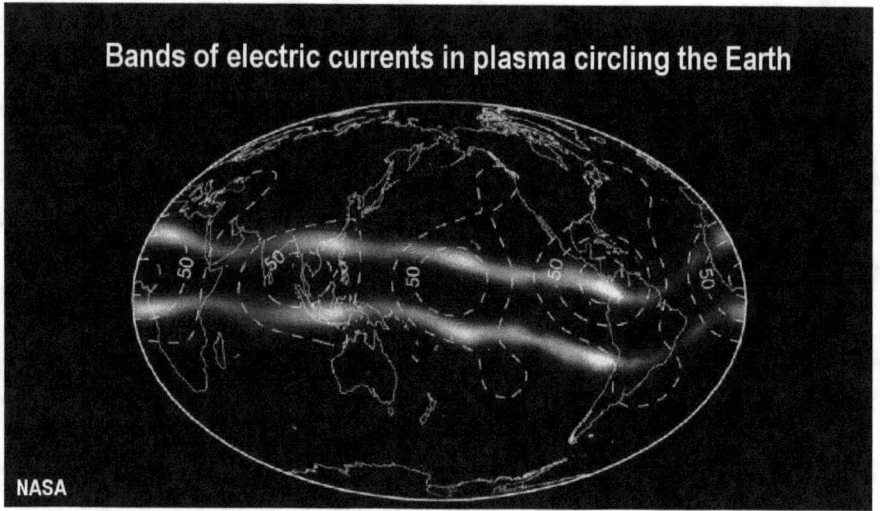

On Earth, a plasma-flow pattern is 'visible' in the ionosphere. The enormous electric potential that is visible there, appears to supply the driving energy for a long list of natural processes on Earth.

The sky is not the limit

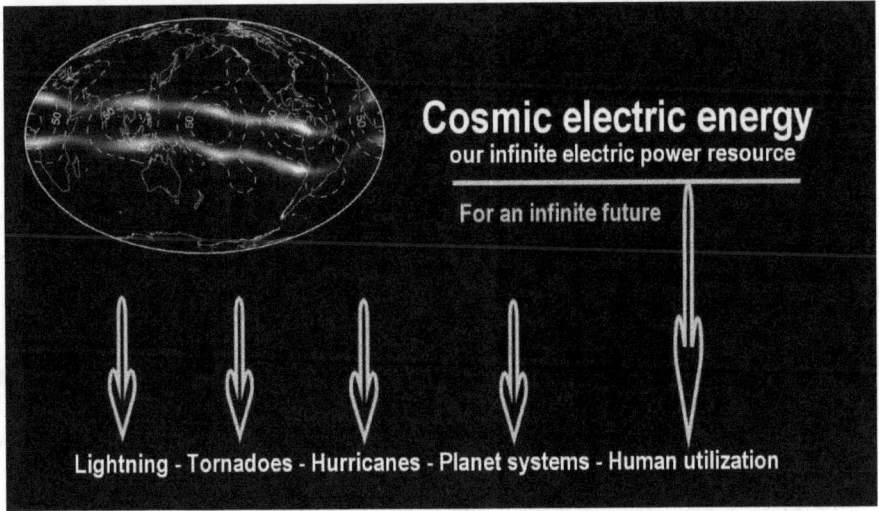

Cosmic electric energy
our infinite electric power resource

For an infinite future

Lightning - Tornadoes - Hurricanes - Planet systems - Human utilization

Among the electric systems powered from the ionosphere are hurricanes, lightning, and tornadoes, whenever thermal events create a sufficiently conductive connection to the ionosphere. We can step up to the plate and participate in the utilization of the cosmic power system, and beyond that, by replicating on Earth the cold fusion principle that the Sun has utilized for its past billions of years. When we get to this point, nuclear fusion, as a synthesizing fusion, will power the economies of humanity. Then the old saying will have a new meaning attached, that "the sky is not the limit." The limit is in the mind. The Sun is not an energy producer, but is an energy converter. The universe doesn't need to produce energy. It is energy. It employs its energy as a creative force.